AF182998

Topology of Surfaces, Knots, and Manifolds

Topology of Surfaces, Knots, and Manifolds

Editor

Johanna Adison

scitus
academics

Topology of Surfaces, Knots, and Manifolds
Edited by **Johanna Adison**

ISBN: 978-1-68117-654-3
Library of Congress Control Number: 2016946521

© 2017 by
SCITUS Academics LLC,
616, Corporate Way, Suite 2, 4766,
Valley Cottage, NY 10989

www.scitusacademics.com

This book contains information obtained from highly regarded resources. Copyright for individual articles remains with the authors as indicated. All chapters are distributed under the terms of the Creative Commons Attribution License, which permits unrestricted use, distribution, and reproduction in any medium, provided the original author and source are credited.

Notice

Reasonable efforts have been made to publish reliable data and views articulated in the chapters are those of the individual contributors, and not necessarily those of the editors or publishers. Editors or publishers are not responsible for the accuracy of the information in the published chapters or consequences of their use. The publisher believes no responsibility for any damage or grievance to the persons or property arising out of the use of any materials, instructions, methods or thoughts in the book. The editors and the publisher have attempted to trace the copyright holders of all material reproduced in this publication and apologize to copyright holders if permission has not been obtained. If any copyright holder has not been acknowledged, please write to us so we may rectify.

Printed in United States of America on Acid Free Paper

Preface

Since the early part of the 20th century both topology and analysis have fed off each other and this has resulted in some very beautiful theorems that lie at the interface between these two disciplines. Perhaps the best known example of these is Brouwer's fixed point theorem (later generalised by J. Schauder to more general spaces) which has had countless applications in applied mathematics, economics and analysis itself.

Topology is the study of those properties of objects that are preserved under careful deformation. Topology is the area of mathematics which investigates continuity and related concepts. Important fundamental notions soon to come are for example open and closed sets, continuity, and homeomorphism. Originally coming from questions in analysis and differential geometry, by now topology permeates mostly every field of math including algebra, combinatorics, logic, and plays a fundamental role in algebraic/arithmetic geometry as we know it today.

Topology of Surfaces, Knots, and Manifolds offers an intuition-based and applied approach to the basic ideas and problems involving manifolds, particularly one- and two-dimensional manifolds. A comprehensive, self-contained treatment presenting general results of the theory.

Table of Contents

CHAPTER 1

Topology and the Visualization of Space

Tony Robbin

Independent Artist; 423 Broome St., New York, NY 10013, USA

ABSTRACT

Overlapping patterns provide the diagrammatics for four-dimensional space. If these patterns are three-dimensional lattices, and if one imagines them extended in three-dimensional space, then the diagram makes a model of physical space.

Keywords: four-dimensional geometry; braiding; projection

1. INTRODUCTION

As Scott Carter has explained to me [1], to braid threads one must go over and under the one-dimensional threads—passing through two more dimensions than the threads themselves. To braid sheets, not ribbons but sheets, one needs access to four spatial dimensions, to go over and under the infinite two-dimensional sheets. In my painting, five fat sheets, each with a different color and a different internal structure, are shown braided together, slipping over and under each other in higher-dimensional space.

2. THE FIVE SHEETS

In the painting reproduced as Figure 1, the five sheets are:

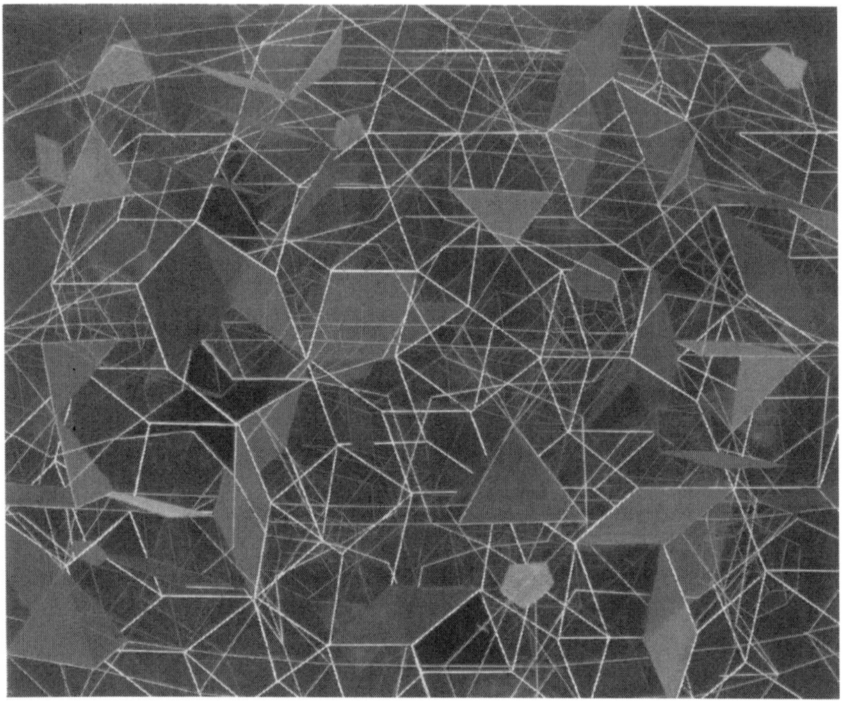

Figure 1. *2006-6*, 2006, Acrylic on Canvas, 56 × 70 in. Collection of the Artist.

Figure 2 is a quasicrystal lattice in 5-fold orientation where the acute angles are 72° and 36°; it is a slice through a quasicrystal cloud that was made with the deBruijn algorithm [2] and see Appendix, where whole cells are taken. (Centers of the cells are computed, and when centers are within certain parameters, whole cells are kept. Software of my design.) This lattice is yellow in the painting.

Figure 2. 5-fold.

I call the pattern in Figure 3 *Persian Dome.* The two fundamental tessellation elements are an elongated octagon, and a four-pointed star, which also has eight sides. The octagons are tessellated along two perpendicular axes, and are slightly separated forming small four-pointed stars as negative shapes. The larger four-pointed stars are tessellated point to point. The pattern was taken from Islamic architecture, and "domed" by me using Formian software (see note at end of essay) from the University of Surrey in Guilford. In the painting, this pattern is light green.

Also created and curved with Formian is the lattice that I call *Nose* (Figure 4). Octahedra are tessellated vertex to vertex in the north-south axis, but they are separated by line segments in the east-west axis. The curvature is on the order of a nosecone, but whether or not it is exactly parabolic is hidden in the Formian software. In the painting, this lattice is light blue.

Figure 3. Persian Dome.

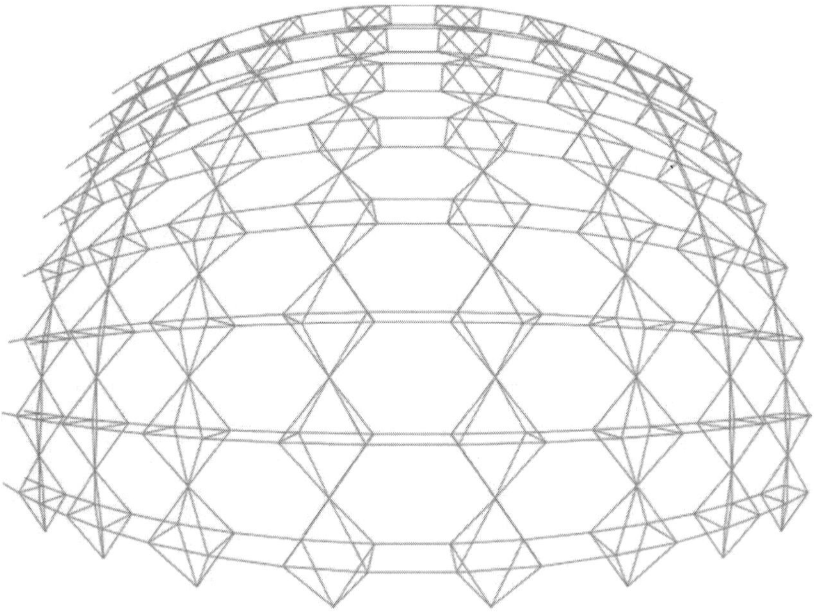

Figure 4. Nose.

Dodecahedra flat (Figure 5), is a disconnected lattice; such lattices are anathema to topologists (how would one count the holes?), but the disconnected lattice suits my purpose, which is to define a slab of space without cluttering the field of view. Formian generated this field of dodecahedra, drawn in perspective, and oriented flat (or almost flat) to the picture plane. The dodecahedra are a delicate white.

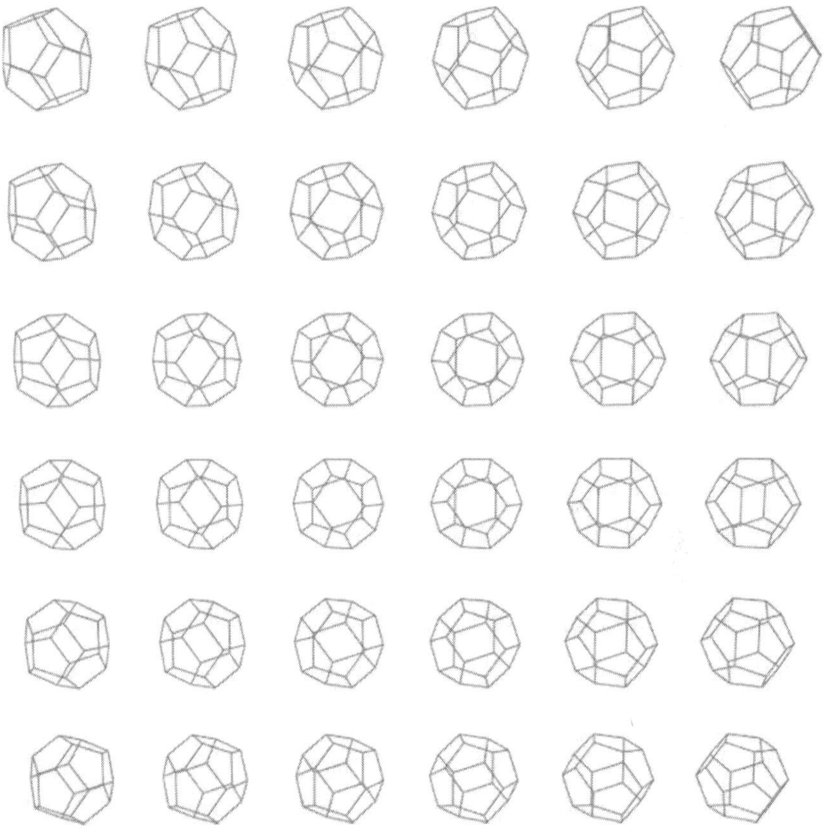

Figure 5. Dodecahedra flat.

In the painting *2ZHP* (Figure 6), is a field of rhombic dodecahedra, which are the simplest sub-assemblies in a three-dimensional quasicrystal. There are three such sub-assemblies, often called zonohedra [3]. Shown above is the complete hypar projection, but only a section of this is used in the artwork. The disconnected vermillion figures dance around, twisting and flexing, as the zonohedra are shown projected to a doubly-curved surface and drawn in perspective.

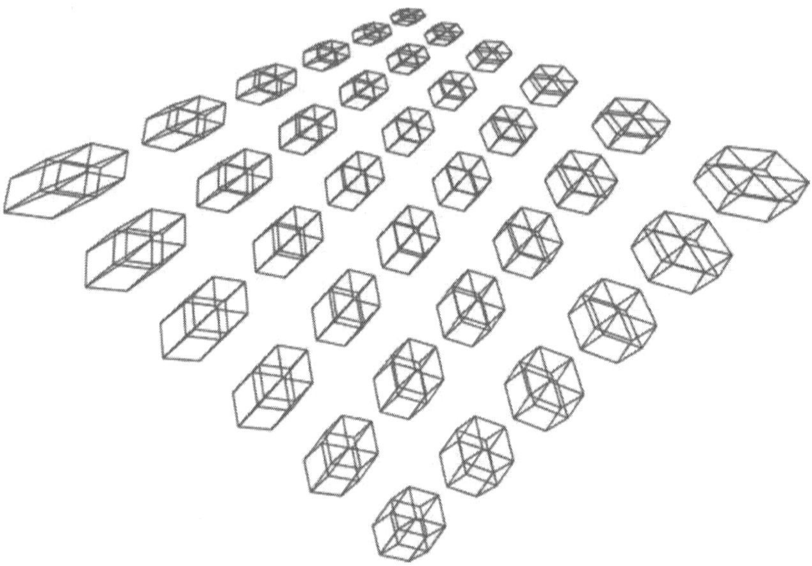

Figure 6. 2ZHP.

3. OVERLAP, BRAIDING AND THE ARTIFACTS OF PROJECTION

Elements in each of the five patterns (let us also call the lattices and disconnected fields *patterns*) are filled in with color, which are sometimes transparent and sometimes opaque. These filled plates define the braiding at that point. For example the vermillion rhombic dodecahedron in the upper left has three plates filled with either red, orange, or yellow. Here, the rhombic dodecahedra pattern is clearly on top of the light yellow 5-fold, but elsewhere in the painting it is behind the 5-fold. Small pentagons that are color-filled plates of the dodecahedra pattern are behind the yellow 5-fold at some locations, yet on top of the 5-fold at other places. There are two complications to the overlapping: objects can be close to the viewer in the third dimension and far away in the fourth, and projection from four to three (and again to a picture plane) can induce apparent, paradoxical intersections.

To accept the first complication, imagine that you are only your eyes looking out a picture window. The mountains are the same distance from the floor as you are—the same Y component. Although close in the *Y* direction, the mountains are far away in the *Z* direction. Likewise, a pattern can remain on top, be consistently close in the third dimension, while parts of it slip far behind in the fourth dimension. Secondly, whereas infinite patterns may slip over and under each other in the $n + 2$ dimension, there are apparent self-intersections as artifacts of the projection when these overlappings are projected to an $n - 1$ dimension.

Sorting out all these complications is not the point. My paintings are not equations, and it is not possible to read a mathematical resolution in them. Rather they revel in the richness and paradox of higher-dimensional visual phenomena. With a knowing nod to the mathematical possibilities, the paintings encourage an acceptance of such spatial complexity. Further, they encourage a taste for, and even a giddy joy, in spatial complexity.

There is a 100 year history of artists being influenced by the "fourth dimension", either the four-dimensional geometry of the platonic solids in dimension 4, or a mysticism rationalized by the fourth dimension, or by considering time as the fourth dimension. This increasingly well-known history is best studied in the many works of Linda Henderson, beginning with *The Fourth Dimension and Non-Euclidean Geometry in Modern Art* [4]. I know of no artist other than myself, however, who has studied and taken inspiration from higher-dimensional topology.

4. FAT SHEETS, FATTER AND FATTER

In 1952, Albert Einstein added a fifth appendix to his popular 1916 book *Relativity: The Special and the General Theory*. The following quote, which I read forty years ago, is one of the most important sources for my art, of which I count only three or four.

"When a smaller box s is situated, relatively at rest, inside the hollow space of the larger box S, then the hollow space of s is a part of the hollow space of S, and the same "space", which contains both of them, belongs to each of the boxes. When s is in motion with respect to S, however, the concept is less simple. One is then inclined to think that s encloses always the same space, but a variable part of the space S. It then becomes necessary to apportion to each box its particular space, ***not thought of as bounded****, and to assume that these two spaces are in motion with respect to each other.*

Before one has become aware of this complication, space appears as an unbounded medium or container in which material objects swim around. But it must now be remembered that there is an infinite number of spaces, which are in motion with respect to each other" [5], pp. 138–139, emphasis added.

At the end of the paragraph, Einstein adds that this notion of "spaces" is "far from having played a considerable role even in scientific thought".

My abbreviated paraphrase of the Einstein quote, *many spaces in the same space*, became my shorthand for a more complicated view of both the objective space of relativity physics and the subjective space of my experience. I played different roles: simultaneously a son and a father. I worked in or was influenced by different traditions: Persian miniatures and Abstract Expressionism; nineteenth century French painting and computer art. I lived in and was formed in different places: Japan and Iran; New England and Europe.

This "spaces" view of special relativity allows us to notice that the multiple separate boxes *s*, even when expanded, are projected into the single larger space *S*. Elsewhere [6,7], I have written of the importance of this observation: the different special relativity formalisms of Henri Poincaré, Hermann Minkowski, and Roger Penrose are necessarily mathematically equivalent, and they can all be seen to be projective systems, most obviously in Penrose's formulation of the Lorentz transformations as projective transformations on the Riemann sphere. Since much of quantum mechanics takes place in a projective space, and since projective space seems necessary for quantum information theory's depiction of non-local phenomenon, here perhaps is a major clue to understanding reality: it is a projective phenomenon. A light ray has a direction, but no fixed length depending on the motion of the observer; it could even be a dimensionless point in its own frame of reference, in its own box traveling at the velocity of light. The light ray is best described as a projective point, rather than a line in space.

We tend to think with the calculus slicing model of space: all of space at T_1, then T_2, then T_3, *etc.* It might actually be more productive to think with the seemingly paradoxical projective model: all of the spaces in the same space at the same time. I have acquired this opinion by looking at my paintings, and perhaps others could too.

5. SYMBOLS, PICTURES, AND DIAGRAMMATICS

When a Catholic sees a sketch, even a crude sketch, of a woman holding a baby and wearing a blue shawl, there is a rush of feeling: compassion, gratitude, humility, and so on. For that woman is surely the Virgin Mary and the child is the Christ. Someone from a different religion might just see a woman and a baby. An artist would be required to go beyond the symbolism and provide the experience of the holy: by imbuing the scene with a golden light or expressing a great tenderness between mother and child. To be effective on its own, symbolism requires that the viewer be learned in the symbols. The apogee of symbolism is the algebraic equation, which—although it embodies a picture of the world—is nevertheless so coded that only those schooled in its methods can experience this. An artist working with the same vision as, say, those who understand the Lorentz transformation equations, causes even the innocent to experience the vertiginous yet lyrical qualities of that space. Diagrammatics, it seems to me, is somewhere in the middle between the symbolic and the pictorial: although it is not the purpose of mathematicians to dwell on the emotive, the hope is to communicate more fully and to reach more than just the initiate, as diagrammatics engages the visual mind and evokes visual intuitions.

6. CONCLUSIONS

There are visionary physicists who believe that everything that is and that happens is so and does so because of a property of space. Einstein believed that gravity exists because space can be curved. Penrose believes that fundamental, massless particles exist because space can be accurately described by projective geometry. David Bohm believed that non-local, quantum reality exists because space can be infinitely smeared over itself. We honor their vision and join this group when we imagine empty space to be rich and fecund.

CONFLICTS OF INTEREST

The author declares no conflict of interest.

APPENDIX

In the case of the three-dimensional quasicrystal, deBruijn's algorithm starts with a pattern, a tessellation of six-dimensional hypercubes. Each six-dimensional hypercube is composed of five-dimensional hypercubes, each of which is composed of four-dimensional hypercubes, each of which, in turn, is composed of regular cubes, set in six-dimensional space. A single one of these regular, right-angle cubes is selected from this dense tessellation and then multiplied by a projection matrix to deliver one of two differently skewed cubes, oriented, and placed in three-dimensional space. With repeated applications of the algorithm, a three-dimensional quasicrystal will be made. The algorithm, then, has two procedures: selection and projection, and the order in which these procedures are employed is irrelevant to the outcome. The projection matrix is a given in the algorithm; it is the collection of rays normal to the faces of a dodecahedron, often called a star of vectors. Think of these rays as the origin of our starting six-dimensional regular tessellation (all mutually perpendicular) now bent by projection into three-dimensional space. What were 90 degree angles are now angles of the irrational number arccosine one over the square root of five, about 63.44 degrees. The devil is in the selection procedure, which is too complicated to discuss in detail here. Nevertheless, I can hint at the selection procedures used.

If one wished to select first then project, a test is devised to first project to a field perpendicular to the final field to see if the component would pass through a spatial gate. If so, keep it and then proceed to project it to its final position. If not, discard. The testing gate is a triacontahedron whose dimensions are a given in the program. (One Japanese mathematician, whose name I have unfortunately lost, poetically called the triacontrahedron gate a crystal ball one peered into to see the future go/no go of the tested element.) In effect, the result of the test is a thin slice of the six-dimensional cubic lattice. This method is sometimes called the Direct Projection Method or the Slice and Project Method. Alternatively, if one wishes to project (in a given way) the entire six-dimensional hypercubic lattice to three-dimensional space first, then the problem is to choose which skewed cubes tessellate that three-dimensional space, and reject all others of the very many cells that interlace with that tessellation. In a process that I correctly programmed, yet that still amazes and mystifies me, one can cycle through the combinations of projected axes, three at a time, and construct one of only two skewed cubes, in the proper orientation, and placed at the correct location in three-dimensional space. This second method, sometimes called the Grand Dual Method, is superior to the Slice and Project Method in that it can render the zonohederal subgroupings that are nested in the quasicrystal.

A note on Formian: Formian was developed by the Engineering Department at the University of Surrey in Guilford. It was created to generate geometry for architectural spaceframes. It not easy to use, and there is no available manual. Nevertheless, it has functions that are very useful for topologists and artists. It generates patterns by reflecting, rotating, and translating simple elements; it has many procedures for curving patterns and lattices; and it allows one to work in an arbitrary number of dimensions. Extra dimensions can be added or subtracted as the patterns are being generated: it is not necessary to specify data structures in advance.

REFERENCES AND NOTES

1. Many thanks to Scott Carter for conversations, emails, and books since 1999
2. De Brujn, N.G. The Algebraic Theory of Penrose's Non-Periodic Tiling of the Plane. *Indag. Math* **1981**, *84*, 39–66.
3. The other zonohedra are the rhombic icosahedron made of five fat cells and five skinny cells, and the rhombic triacontrahedron, made of ten cells of each kind. Sometimes the single fat cell is also called one of the zonohedra.
4. Henderson, L.D. *The Fourth Dimension and Non-Euclidean Geometry in Modern Art*; Princeton University Press: Princeton, NJ, USA, 1983.
5. Einstein, A. Relativity and the Problem of Space. In *Relativity: The Special and the General Theory*, 15th ed.; Crown Publishers: New York, NY, USA, 1952.
6. Robbin, T. *Shadows of Reality, the Fourth Dimension in Relativity, Cubism, and Modern Thought*; Yale University Press: New Haven, CT, USA, 2006.
7. Robbin, T.; David, Bohm; Roger, Penrose. *the Search for Non-local Causality*, Available online: http://www.interaliamag.org/articles/david-bohm-roger-penrose-and-the-search-for-non-local-causality/ accessed on 22 December 2014

2

CHAPTER

General Topology of the Universe

Aalok Pandya[1,2]

[1]*Department of Physics, University of Rajasthan, Jaipur, India*

[2]*Department of Physics, JECRC University, Jaipur, India*

ABSTRACT

General topology of the universe is described. It is concluded that topology of the present universe is greater or stronger than the topology of the universe in the past and topology of the future universe will be stronger or greater than the present topology of the universe. Consequently, the universe remains unbounded. The general topological approach comprises of powerful techniques that could prove to be useful to prescribe mathematical constraints on the global character of the universe as well as on the manifold of space-time.

Keywords: General Topology, Manifold, Boundary, Causality, Euclidean Space, Discreteness, Limit Points

1. INTRODUCTION

General topology of the universe and its evolutionary aspects are discussed in this paper. In the scenario of universe evolving with time, we conclude that topology of the present universe is greater or stronger than the topology of the universe in the past and topology of the future universe will be stronger or greater than the present topology of the universe. Consequently, the universe remains unbounded. The conditions of whole space topology can be treated as mathematical constraints on the global topology of the universe. The present exercise draws motivation from the recent explorations [1] [2] of relativistic aspects of the causal evolution of world lines and the earlier work on causal set topology by Joshi [3] . The Null-Cone representation by means of causal set topology [1] [2] revives interests in topological analysis of the universe. In earlier attempts, topological aspects of the large-scale structures in the universe have been discussed by Hartle and Hawking [4] , Ellis and Hawking and others [4] - [6] . Topological aspects of the birth of the universe discussed by Miao Li [7] also lead to a substantial development of mathematical ideas on physical topology of the Universe. The geometry and topology of the universe in the physical realm have been discussed by Lachieze-Rey and Luminet [6] and Mellot [8] . All the care is taken to ensure that fundamental concepts of general topology are not violated. Moreover, the present discussion has been developed with a focus on the earlier studies such as Pankaj Joshi [3] , Penrose [9] and Vyas [10] wherein causal structure and general topology of the space-time and the global aspects in gravitation and cosmology were discussed rigorously. The computer simulation and animation in $3+1\ D$ space in the reference [2] motivates us to further investigate the unexplored aspects of the general topology of space-time.

It is felt that apart from the manifold of the universe, other general topological properties of the universe and the topological changes following them are equally important. The general topology of the universe can be uniquely described if its physical features are specified. A topological space is an abstract conception and thus it is always not possible to attribute physical properties to it. But, the other way round is certainly possible. The topology of a physical object with evident metricity that is inbuilt distance features can certainly be described. Moreover, it is pertinent to mention that the physical realities of the universe [3] - [8] are in perfect agreement with the mathematical conditions discussed in the present paper. Thus, it is good ground to assume that topological considerations could prove to be crucial mathematical constraints on the overall geometry of the universe.

We have several models of the universe at the cross-roads of scientific development, but we consider only one scenario wherein the universe is the single largest entity that encloses every physical object that exists in nature including space and time. The discussion in this paper is developed on two logical grounds such that firstly it justifies the observed facts and secondly it satisfies the mathematical conditions of the whole space topology. The universe without boundary has already been described in the framework of General Theory of Relativity (GTR) too. But, we feel that GTR and Einstein equation describe the universe in a specific physical scenario of a theoretical framework, whereas, general topology of the whole space is absolute truth, and thus it is not limited to any specific physical scenario. This discussion addresses not only the boundary of the universe, but also a detail spatial analysis of the universe with reference to open space, closed space, boundary, compactness, connectedness, limit points, frontier points, interior points, topological transitions and causality and topology. It is concluded in this paper that the general topology of the whole space could be a very good philosophical basis for describing geometry of the universe as it prescribes mathematical constraints on the global geometry of the universe.

2. OPENNESS, COMPACTNESS, AND THE BOUNDARY

The metric of the universe can be considered as the super metric or the universal metric defined on the whole space U. Then such a universal topology is the set of all possible sets and subsets including the null set. With the consideration that everything that exists in nature is contained in the universe, the complement of the universe is obviously nonexistent.

By definition universal topology being collection of all possible sets and subsets, is always discrete in nature. Thus, the topology of the universe is a discrete topology. Also, the whole space U is always open as well as closed. We explain this geometrical duality using both geometrical physical explanations. Since, $U \cap \{\phi\} = \{\phi\}$; U is closed if its complement $\sim U$ is open and U is open if its complement $\sim U$ is closed, as the null space ϕ is both open as well as closed.

Mathematically, if the closure of the space U is given by $\bar{U}$, the condition of the whole space implies that: $U = \bar{U}$.

Hence, if the closure of the universe is equal to the universe itself which is precisely true for the physical universe, the set U of the universe is said to be closed.

An important mathematical corollary [11] asserts that boundary of a space can be given by: $\text{Bd } U = \bar{U} - U$.

But, since $U = \bar{U}$, the boundary of the space vanishes as $\text{Bd } U = \phi$.

Thus, $\text{Bd } U = \phi \Leftrightarrow$ is both open and closed.

This agrees with the conclusion which we have already drawn but by another reason. It is interesting to cite here a mathematical theorem, which states that boundary of a boundary is always empty [11] :

$$\partial\partial M = \phi.$$

Furthermore, if d is a metric defined on space U, a subset A of U is said to be bounded if there is some number N such that

$d(a_1, a_2) \leq N$; for every pair of points a_1, a_2 of A.

If A is bounded, the diameter of A is defined to be the number

$$\text{diam}(A) = \text{least upper bound } \{d(a_1, a_2) : a_1, a_2 \in A\}.$$

But, we know that for continuously expanding U, particularly in inflationary scenario, this condition cannot be fulfilled as for every pair a_1, a_2 ; the distance $d(a_1, a_2)$ cannot remain less than or equal to a fixed number N forever. The aforesaid conclusion is in agreement with the statement that the universe has no boundary, because a boundary represents in a sense the "edge" of the universe not detected by any astronomical observations [4] [6] . However, the metric of the universe is closed but it cannot be bounded, and therefore it is not at all compact.

3. CAUSALITY AND SPACE-TIME TOPOLOGY

An event p chronologically precedes another event q, denoted by $p \ll q$, if there is a smooth future directed time-like curve from p to q [1] - [3] [9] [10] .

If such a curve is non-space-like that is, either time-like or null, we say that p causally precedes q, or $p < q$. The chronological future $I^+(p)$ of p is the set of all points q such that $p \ll q$. The chronological past of p is defined dually [3] . Thus we have

$$I^+(p) = \{q \in M \mid p \ll q\}, \tag{1}$$

And

$$I^-(p) = \{q \in M \mid q \ll p\}. \tag{2}$$

The causal future (past) for p can be defined similarly

$$J^+(p) = \{q \in M \mid p < q\}, \tag{3}$$

$$J^-(p) = \{q \in M \,|\, q < p\}. \tag{4}$$

We now observe yet another duality, which can be realized for all $p \ll q \ll r$, such that

$$I^-(r) \supset I^-(p), \tag{5}$$

but,

$$I^+(r) \subset I^+(p). \tag{6}$$

And similarly, causal future (past) could be defined dually as:

$$J^-(r) \supset J^-(p), \tag{7}$$

And

$$J^+(r) \subset J^+(p). \tag{8}$$

Proposition 3.1: If there is a smooth future directed time-like curve with all $p \ll q \ll r$ on it, then

$I^-(r) \supset I^-(p)$ and $I^+(r) \subset I^+(p)$.

Proof: If we consider $p \ll q \ll r$ in the given case, then obviously

$q \notin I^+(r)$, and $q \in I^+(p) - I^+(r)$.

Thus, the fact that $I^+(p) - I^+(r) \neq \phi$ implies $I^+(r) \subset I^+(p)$.

Similarly, one can easily assert that: $q \notin I^-(p)$ and $q \in I^-(r) - I^-(p)$.

And, the fact that $I^-(r) - I^-(p) \neq \phi$ implies $I^-(p) \subset I^-(r)$.

Hence is the proposition.

The computer simulation containing animation in $3+1\ D$ space in the reference [2] is reassuring that this discussion is on firm footing.

Also, this conveys the feature of open-ended future of the universe at any point of time. If the universe goes on expanding then the topology of the universe will tend to be greater or stronger topology. And the topology of the universe at any point of time in future will be stronger or greater than the present topology, and it will further tend to be greater and stronger. This is due to the fact that it will have more and more number of open sets, following this enhancement.

Proposition 3.2: The topology of the present universe is greater or stronger than the topology of the universe in the past and topology of the future universe will be stronger or greater than the present topology of the universe.

Proof: If we consider the topology of the universe defined on the whole space $(U,\mathfrak{I})$, at some point of time in the past then we can construct the topologies of the universe at present and in the future, relative to this. The topology of the present universe can be given as:

$$\mathfrak{I}_{\text{present}} = \{U, \phi, \cup\{x_\mu\}\};$$

where, $\cup\{x_\mu\} \subset U$ is the collection of all space-time points accumulated in U, after having defined the topology $\mathfrak{I}_{\text{past}}$.

Similarly, we can define the topology of the universe of future as:

$\mathfrak{I}_{\text{future}} = \{U, \phi, \cup\{x_\nu\}\}$, where $\cup\{x_\nu\} \subset U$, and $\cup\{x_\nu\}$ is the collection of all space-time points accumulated in U, after having defined the topology $\mathfrak{I}_{\text{past}}$; such that $\cup\{x_\nu\} \supset \cup\{x_\mu\}$.

Hence, $\mathfrak{I}_{\text{past}} \subset \mathfrak{I}_{\text{present}} \subset \mathfrak{I}_{\text{future}}$.

In other words the topology of $U(\text{past})$ is smaller or weaker than the topology of $U(\text{present})$, and topology of $U(\text{future})$ will be greater or stronger than the topology of $U(\text{present})$.

Hence is the proposition.

This further guarantees that the universe will remain unbounded. We can alternatively explain the topological transition of the universe from a weaker or smaller to a stronger or greater topology in terms of matter-energy distribution. Considering the total energy of the universe to be conserved, the continuous expansion of the universe will lead to increase in the intergalactic distances and decline in the matter density over the space. This implies that the topology of the universe in past $U(\text{past})$ was coarser than the $U(\text{present})$, and topology of $U(\text{future})$ will be finer than the $U(\text{present})$.

4. LIMIT POINTS

Now, we look for the limit points of the universe. However, this part of the discussion may not be applicable to the consideration of multi-universe Cosmos. If the universe is not static and is continuously expanding, its farthest possible extents can be easily approximated. The radius of the expanding universe can be given by the

well-known relation $R = \frac{c}{H}$, or alternatively $R \approx c\tau$; where H^{-1} is Hubble age, c is the velocity of light,

and τ is characteristic time scale or the age of the universe. Alternatively, for the sake of generality, this distance L_{Horizon} that is (comoving) radius of the observable universe (with or without inflation) is given with a good precision as [6] :

$$L_{\text{Horizon}} \approx \frac{u}{(1-\gamma)} L_{\text{Hubble}}.$$

This follows from the logic that the size of the universe can be determined by means of the distance traversed by the fastest moving particle ever since the genesis of the universe. Obviously, this radius is only instantaneous radius and is increasing continuously. Moreover, the universe is considered to be homogeneous and isotropic. With the expansion of the universe, the continuous inclusion of more number of space (or space-time) points restricts the universe from being bounded and compact.

5. DISCUSSION

One can define a manifold over a topology if that topological space is locally homeomorphic to R^n, which means it is Hausdorff and possesses metric properties, and connectedness is well described in that. Moreover, a smooth manifold can be described only if continuity as well as differentiability is well defined on it. The manifold of the universe can only be described, if it satisfies these requisite conditions. And we know that it does satisfy them.

Also, if the origin of the universe is considered to have followed the single event say Big-Bang, then all the physical objects and events in the universe are causally connected. Thus, one can say that the metric of the space-time is connected too. Physically interesting known space-time examples such as the Schwarzschild geometry and Robertson-Walker space-times are topologically Hausdorff. Also, for the sake of the causal regularity of space-time, the continuum of the space-time structure is necessary which would ensure a globally well- behaved space-time [3] . In nut-shell, at least for the Minkowski space-time, the manifold is globally Euclidean with the topology of R^4 over the Lorentzian metric. In case of general theory of relativity, the principles of GTR effectively imply that it is the space-time metric, and the quantities derived from it, that must appear in the equations for physical quantities and that these equations must reduce to the flat space-time case when the metric is Minkowskian. This is the basic content of the GTR where the space-time manifold is allowed to have topologies other than R^4 too. However, locally and globally, there could be important differences in the causal structure due to a different local space-time topology, strong gravitational fields and so on [3] .

There is a reason to extend the understanding of topology of space-time to the universe at large. Once we understand the nitty-gritty of causal structure and topological aspects of space-time, the understanding of causal structure enables us to deliberate space-like, light-like and time-like events. And the universe at large is nothing but an assembly of space-like, light-like and time-like events. However, the manifold of the universe can have different possible topological features depending upon the physical conditions considered in a specific scenario the global aspects of the general topology of the universe as the whole space remain unchanged. Thus, a prominent suggestion emerges that the mathematical conditionality of the whole space topology can be used as a constraint to describe the overall topology of the universe. In other words, the general topological approach comprises of powerful techniques that could prove to be useful to prescribe mathematical constraints on the global character of the Universe as well as on the manifold of the space-time. In future we can explore frontier points, interior points and other topological notions too.

ACKNOWLEDGEMENTS

The author is grateful to Professor T. Pandmanabhan of Inter University Centre for Astronomy and Astrophysics (IUCAA) for his encouragement and patient reading of the manuscript.

REFERENCES

1. Aalok, P. (2012) Null Cone Representation of Interactions and Causal Structure. Physics Essays, 25, 228-232.

2. Aalok, P. (2013) Computational Investigations of the Null Cone Representation of Interactions. Physics Essays, 26, 24-26.

3. Joshi, P.S. (1993) Global Aspects in Gravitation and Cosmology. Oxford Science Publications, Oxford.

4. Hartle, J.B. and Hawking, S.W. (1983) Wave Function of the Universe. Physical Review D, 28, 2960-2975; Hawking, S.W. (1984) Quantum State of the Universe. Nuclear Physics B, 39, 257-276.

5. Hawking, S.W. and Ellis, G.F.R. (1973) The Large Scale Structure of Space-Time. Cambridge University Press, Cambridge.

6. Lachieze-Rey, M. and Luminet, J.P. (1995) Cosmic Topology. Physics Reports, 254, 135-214.

7. Li, M. (1986) On Topological Aspects of the Birth of the Universe. Physics Letters B, 173, 229-232.

8. Melott, A.L. (1990) The Topology of the Large Scale Structure in the Universe. Physics Reports, 193, 1-39.

9. Penrose, R. (1972) Techniques of Differential Topology in Relativity. A. M. S. Colloquium Publications, SIAM Philadelphia.

10. Vyas, U.D. and Joshi, P.S. (1989) Topological Techniques in General Relativity in Geometry and Topology. World Scientific, Singapore, 315-331.

11. Munkres, J.R. (1975) Topology: A First Course. Prentice-Hall Inc., Englewood Cliffs.

3
CHAPTER

Fixed Points of Local Actions of Lie Groups on Real and Complex 2-Manifolds

Morris W. Hirsch [1,2]

[1]*Department of Mathematics, University of Wisconsin, Madison, WI 53706-1388, USA*

[2]*Department of Mathematics, University of California, Berkeley, CA 94720-3840, USA*

ABSTRACT

I discuss old and new results on fixed points of local actions by Lie groups *G* on real and complex 2-manifolds, and zero sets of Lie algebras of vector fields. Results of E. Lima, J. Plante and C. Bonatti are reviewed.

Keywords: Lie group; Lie algebra; vector fields; fixed point index; dynamical systems

1. INTRODUCTION

Classical results of Poincaré [1] (1885), Hopf [2] (1925) and Lefschetz [3] (1937) yield the archetypal fixed point theorem for Lie group actions:

Theorem 1.

Every flow on a compact manifold of non-zero Euler characteristic has a fixed point.

Here the Lie group is the group of real numbers.

The earliest papers I have found on fixed points for actions of other non-discrete Lie group are those of P. A. Smith [4] (1942) and H. Wang [5] (1952). Then came Armand Borel's landmark paper of 1956:

Theorem 2

(Borel [6]). *If H is a solvable, irreducible affine algebraic group over an algebraically closed field* KK, *every algebraic action of H on a complete algebraic variety over* KK *has a fixed point.*

Over the field of complex numbers, completeness is equivalent to compactness in the classical topology, and complete nonsingular varieties are compact Kähler manifolds.

In 1973, A. Sommese [7] extended Borel's theorem to solvable holomorphic actions on compact Kähler manifolds with first Betti number 0. In contrast to the results below, these have no explicit restrictions on dimensions or Euler characteristics.

2. ACTIONS AND LOCAL ACTIONS

If f:A→B denotes a map, its domain is Df:=A and its range is Rf:=f(A)R.

Let g,fg,f denote maps. Regardless of their domains and ranges, the composition g∘fg∘f is defined as the map x↦g(f(x))whose domain, perhaps empty, is f−1(Dg). The associative law holds for these compositions: The maps (h∘g)∘f and h∘(g∘f) have the same domain

$$D := \{x \in \mathcal{D}f:\ f(x) \in \mathcal{D}g, \quad g(f(x)) \in \mathcal{D}f\},$$

and

$$x \in D \implies (h \circ g)(f(x)) = h((g \circ f)(x)).$$

Henceforth *M* denotes a manifold with boundary *∂M*, and *G* denotes a connected Lie group with Lie algebra $\mathfrak{g}$.

A *local homeomorphism f* on *M* is a homeomorphism between open subsets of *M*. The set of these homeomorphisms is denoted by LH(M)LH(M).

A *local action* of *G* on *M* is a triple (α,G,M), where α:G→LH(M) is a function having the following properties:

- The set $\Omega(\alpha) := \{(g, p) \in G \times M:\ p \in \mathcal{D}\alpha(g)\}$ is an open neighborhood of {eG}×M{eG}×M.
- The *evaluation map* $\mathrm{ev}_\alpha:\ \Omega(\alpha) \to M, \qquad (g, p) \mapsto \alpha(g) \cdot p$ is continuous.
- α(eG) is the identity map of *M*.
- The maps α(fg)∘α(h) and α(f)∘α(gh) agree on the intersection of their domains.
- α(g−1)=α(g)−1.

Notation of *α* may be omitted.

When Ω(α)=G×M the local action is a *global action*. If *G* is simply connected and *M* is compact, every local action extends to a unique global action.

When *α* has been specified, we define the *fixed-point sets*

$$\mathrm{Fix}(g) := \{x \in \mathcal{D}g\colon\ g(x) = x\},$$
$$\mathrm{Fix}(G) := \bigcap_{g \in G} \mathrm{Fix}(g)$$

The local action is *effective* if Fix(g)≠M for all g≠eG.

A *local flow* is a local action (Ψ,R,M). In this case we set Ψt:=Ψ(t) and identify Ψ with the indexed family of {Ψt}t∈R of local maps in *M*. If (α,G,M) is a local action, to every $\Phi^X := \{\Phi_t^X\}_{t\in\mathbb{R}}.$ there corresponds a local flow $(\alpha^*\Psi,\ R,\ M\ )$ defined in the following. Consider *X* as a 1-parameter subgroup of *G*, i.e., a homomorphism $X\colon\ R \to G$, and set $\alpha^*\Psi = \{\alpha(X(t))\}_{t\in R}$. The local flow induced by a C^1 vector field *X* on *M* tangent to ∂M is denoted by $\Phi^X := \{\Phi^X\}_{t\in R}$.

A *block* for a local flow Ψ (a Ψ-block) is a compact $K \subset \mathrm{Fix}(\Psi)$ having a precompact open neighborhood $U \subset M$, termed *isolating*, such that $\mathrm{Fix}(\Psi) \cap U = K$. When this holds, the *index* $i(\Psi, U)$ of Ψ in *U* is defined as the fixed point index of $\Psi_t|U\colon\ U \to M$ for sufficiently small $t > 0$, as defined by Dold [8] (see also Brown [9] and Granas and Dugundji [10]). This integer depends only on *K*, and we set $i_K(\Psi) := i(\Psi, U)$. When $i_K(\Psi)$ 6= 0 then *K* is *essential*. If *K* is a block for the local flow Φ^X of a vector field *X*, an equivalent definition of $i_K(\Phi^X)$ as the Poincaré–Hopf index of *X* at *K* is given in Section 4.

3. FIXED POINTS OF LOCAL ACTIONS ON SURFACES

In the rest of this section *M* denotes a real closed surface (compact with empty boundary) and *G* is a connected Lie group acting continuously on *M*.

An important role is played by the group ST∘(n,R)ST∘(n,R), the solvable group of real, upper triangulable n×nn×n matrices with positive diagonal entries. In his pioneering 1964 paper, E. Lima

[11] constructed fixed-point free actions of ST∘(2,R)ST∘(2,R) on the compact 2-cell and the 2-sphere, but he also showed that every abelian Lie group action on a compact surface *M* of nonzero Euler characteristic $\chi(M)\chi(M)$ has a fixed point. These results were extended in 1986 by Plante:

Theorem 3

(Plante [12]). *Let M be a compact surface whose boundary may be nonempty.*

(i) ST∘(2,)ST∘(2,) *has a fixed-point free action on M.*

(ii) *If* $\chi(M)\neq 0\chi(M)\neq 0$, *every action on M by a connected nilpotent Lie group has a fixed point.*

Many facts about existence of fixed points for continuous actions on closed surfaces can be derived from the results of M. Belliart summarized in the following theorem. If $H\subset GL(n,F)$ denotes a group of matrices, PHPH denotes the quotient of*H* by its center.

Theorem 4

(Belliart [13]). *There is a fixed-point free action of G on M iff one of the following conditions* (a), (b), (c) *holds:*

(a) $\chi(M)>0$ *and G is solvable but not nilpotent.*

(b) $\chi(M)<0$ *and G has* ST∘(2,R) *as a quotient.*

(c) $\chi(M)\geq 0$, *G is semisimple, and either:*

(i) *G has* PSL(2,R) *as a quotient, or*

(ii) $\chi(M)>0$, $\partial M=\emptyset$, *and G has as a quotient one of the groups*

$$PSL(3,R), PSL(2,C) \text{ or } PSO(3).$$

A Lie algebra is *supersolvable* if it is faithfully represented as upper triangular real matrices. A Lie group is supersolvable if its Lie algebra is.

Theorem 5.

(i) ST∘(3R,) *has an effective analytic action on M.*

(ii) If G has an effective, fixed-point free analytic action on M, then $\chi(M) \geq 0$, with equality when G is a supersolvable and $\partial M = \varnothing$.

Part (i) and the first conclusion in (ii) are due to Turiel [14]. The second conclusion in (ii) is due to Hirsch and Weinstein [15].

The following result gives upper and lower bounds on the number of fixed points of analytic actions of ST∘(3,R):

Proposition 1

(Hirsch [16], Cor. 17, Thm 22).

(i) Let M have genus g. For every k∈Nk∈N there is an effective analytic action β of ST∘(3,R) on M such that:

$$\#\,\mathsf{Fix}(\beta) = \begin{cases} 2(g+k+1) & \textit{if } M \textit{ is orientable,} \\ g+k & \textit{if } M \textit{ is nonorientable and } g \geq 1. \end{cases} \tag{1}$$

(ii) If G is not supersolvable and has an effective analytic action on M,

$$0 \leq \#\,\mathsf{Fix}(G) \leq \chi(M) \leq 2.$$

Question.

Can the right hand side of Equation (1) can be lowered?

4. INDICES OF VECTOR FIELDS

Let V(M) denote the vector space of vector fields (continuous sections of the tangent bundle) on a smooth manifold M, endowed with the compact open topology.

The zero set of X∈V(M) is

$$Z(X) := \{p \in M: \ X_p = 0\}.$$

A block for X (an X-block) is a compact, relatively open set K⊂Z(X). Every sufficiently small open neighborhood U⊂M of K is isolating for X, meaning its closure U¯ is compact and Z(X)∩U¯=K. This implies that U is isolating for every vector field Y sufficiently close to X.

Let K be an X-block. When K is finite, the Poincaré–Hopf index of X at K, and in U, is the integer $i_K^{PH}(X) = \text{i}^{PH}(X, U)$ defined as follows. For each p∈K choose an open set W⊂U meeting K only at p, such that W is the domain of a C1C1chart

$$\phi\colon \ W \approx W' \subset \mathbb{R}^n, \quad \phi(p) = p'.$$

The transform of X by ϕ is

$$X' := T\phi \circ X \circ \phi^{-1} \in \mathcal{V}(W').$$

There is a unique map of pairs

$$F_p\colon \ (W', 0) \to \mathbb{R}^n, 0)$$

that expresses X′X′ by the formula

$$X'_x = (x, F_p(x)) \in \{x\} \times \mathbb{R}^n, \qquad (x \in W').$$

Noting that $F^{-1}(0) = p$, we define $\text{i}_p^{PH}(X) \in \mathbb{Z}$ as the degree of the map defined for any sufficiently small ∊>0 as

$$\mathbf{S}^{n-1} \to \mathbf{S}^{n-1}, \quad u \mapsto \frac{F_p(\epsilon u)}{\|F_p(\epsilon u)\|}$$

where ‖·‖‖·‖ is the norm defined by any Riemannian metric on M. This degree is independent of ∊ and the chart ϕ, by standard properties of the degree function. Therefore the integer

$$\mathrm{i}_K^{PH}(X) = i^{PH}(X,U) := \begin{cases} \sum_{p\in K} i_p^{PH}(X) & \text{if } K \neq \varnothing, \\ 0 & \text{if } K = \varnothing. \end{cases}$$

is well defined and depends only on X and K.

The index of an arbitrary X-block K is the integer iK(X):=i(X,U) defined as the Poincaré–Hopf index of any sufficiently close approximation to X having only finitely many zeros in U [17].

This number is independent of U and is stable under perturbations of X. The X-block K is essential when iK(X)≠0. This implies Z(X)∩K≠∅ because every isolating neighborhood of K meets Z(X).

Theorem 6

(Poincaré–Hopf). If M is compact, i(X,M)=χ(M) for all continuous vector fields X on M.

For calculations of the index in more general settings see Morse [18], Pugh [19], Gottlieb [20], Jubin [21].

Theorem 7

(Bonatti [22]). Assume M is a real manifold of dimension ≤4 with empty boundary, and X,Y are analytic vector fields on M such that [X,Y]=0. Then Z(Y) meets every essential X-block [23].

This implies certain local actions of 2-dimensional abelian Lie groups have fixed points. The results below are analogs for local actions of nonabelian Lie groups.

Theorem 8

(Hirsch [24]). Let M be a real surface, perhaps non-compact or having non-empty boundary. Let G be a connected nilpotent Lie group and (α,G,M) an effective local action. Assume given a continuous local action of G on M, and let K be an essential block for the local flow induced by a 1-parameter subgroup. Then Fix(G)∩K≠∅.

This implies Plante's result, Theorem 3(ii).

Corollary 1.

Let G,M and X be as in Theorem 8.

(i) If Γ⊂M is a compact attractor for Φ^X *and* $\chi(\Gamma) \neq 0$, *then* $\mathrm{Fix}(G) \cap \Gamma \neq \varnothing$.

(ii) If ΦX has n essential blocks, then Fix(G) has n components.

The counter-example in Theorem 3(i) show that fixed point results for broader classes of Lie groups, including supersolvable groups, need stronger hypotheses.

Henceforth M denotes either a real or complex 2-manifold, the corresponding ground field being F=RF=R or CC. Let Vω(M) denote the Lie algebra of vector fields on M that are analytic over FF. If Y∈Vω(M), TΦY denotes the induced local flow on the tangent vector bundle of M.

Assume X,Y∈Vω(M). We say that Y tracks X if there exists a continuous map

$$f\colon M \to \mathbb{F}, \quad f^{-1}(0) = \mathsf{Z}(X), \quad [Y, X] = fX.$$

Equivalently: if p∈Mp∈M and t∈Rt∈R there exists g(t,p)∈F such that:

$$\Phi_t^Y(p) = q(t) \implies T\Phi_t^Y(X_p) = g(t,p)X_{q(t)}.$$

For real M this means ΦYtΦtY sends orbits of $X|\mathcal{D}\Phi_t^Y$ to orbits of $X|\mathcal{R}\Phi_t^Y$.

Let G⊂Vω(M) denote a Lie algebra of vector fields. We say that G$\mathcal{G}$ tracks X provided each Y∈G tracks X.

Example 1.

If X spans an ideal in $G\mathcal{G}$ then $G\mathcal{G}$ tracks X, and the converse holds if $G\mathcal{G}$ is finite dimensional.

Example 2.

The set $\{Y \in V\omega(M)\text{: YtracksX}\}$ is a Lie algebra that tracks X.

The following result will be proved in a forthcoming paper [25]; a preliminary version is in [26].

Theorem 9.

Assume $X \in V\omega(M)$, K is an essential X-bloc, and $G \subset V\omega(M)$ tracks X. Let one of the following conditions hold:

(a) M is complex,

(b) M is real and GG is supersolvable.

Then $Z(G) \cap K \neq \varnothing$.

Example 3.

Here is a simple example in which the hypotheses hold. For M take complex projective 3-space. Let G be the solvable complex Lie group of unimodular 4× 4 upper triangular complex matrices. The natural action of G on C4C4 induces an effective holomorphic action of G on M, mapping the Lie algebra of G isomorphically onto a Lie algebra $G \subset V\omega(M)$. Let $X \in G$ have the block [0010]0100 in its upper right hand corner and all other elements equal to zero. X spans an ideal, the triple commutator subalgebra $G'''G'''$. The X-block $K := Z(X)$, a copy of CP1, is essential because $\chi(M)=3\chi(M)=3$; and Z(G) is a singleton in Z(X).

CONFLICTS OF INTEREST

The author declares no conflict of interest.

REFERENCES AND NOTES

1. Poincaré, H. Sur les courbes définies par une équation différentielle. J. Math. Pures Appl. **1885**, 1, 167–244.
2. Hopf, H. Vektorfelder in Mannifgfaltigkeiten. Math. Ann. **1925**, 95, 340–367.
3. Lefschetz, S. On the fixed point formula. Ann. Math. **1937**, 38, 819–822.
4. Smith, P.A. Stationary points of transformation groups. Proc. Natl. Acad. Sci. USA **1942**, 28, 293–297.
5. Wang, H.-C. A remark on transformation groups leaving fixed an end point. Proc. Am. Math. Soc. **1952**, 3, 548–549
6. Borel, A. Groupes linéaires algébriques. Ann. Math. **1956**, 64, 20–80.
7. Sommese, A. Borel's fixed point theorem for Kaehler manifolds and an application. Proc. Am. Math. Soc. **1973**, 41, 51–54.
8. Dold, A. Lectures on Algebraic Topology. In Die Grundlehren der Matematischen Wissenschaften Bd. 52, 2nd ed.; Springer-Verlag: New York, NY, USA, 1972.
9. Brown, R. The Lefschetz fixed point theorem; Scott, Foresman & Co.: Glenview, IL, USA, 1971.
10. Granas, A.; Dugundji, J. Fixed Point Theory; Springer-Verlag: New York, NY, USA, 2003.
11. Lima, E. Common singularities of commuting vector fields on 2-manifolds. Comment. Math. Helv. **1964**, 39, 97–110.
12. Plante, J. Fixed points of Lie group actions on surfaces. Ergod. Theory Dyn. Syst. **1986**, 6, 149–161.
13. Belliart, M. Actions sans points fixes sur les surfaces compactes. Math. Z. **1997**, 225, 453–465.

14. Turiel, F.-J. Analytic actions on compact surfaces and fixed points. Manuscr. Math. **2003**, 110, 195–201.

15. Hirsch, M.; Weinstein, A. Fixed points of analytic actions of supersoluble Lie groups on compact surfaces. Ergod. Theory Dyn. Syst. **2001**, 21, 1783–1787.

16. Hirsch, M. Actions of Lie groups and Lie algebras on manifolds. In A Celebration of the Mathematical Legacy of Raoul Bott; Amer. Math. Soc.: Providence, RI, USA, 2010.

17. Equivalently: i(X,U) is the intersection number of X|U with the zero section of the tangent bundle (Bonatti [22]).

18. Morse, M. Singular Points of Vector Fields Under General Boundary Conditions. Am. J. Math. **1929**, 52, 165–178.

19. Pugh, C. A generalized Poincaré index formula. Topology **1968**, 7, 217–226.

20. Gottlieb, D. A de Moivre like formula for fixed point theory. In Fixed Point Theory and its Applications (Berkeley, CA, 1986); Amer. Math. Soc.: Providence, RI, USA, 1988.

21. Jubin, B. A generalized Poincaré–Hopf index theorem. arxiv:0903.0697. 2009.

22. Bonatti, C. Champs de vecteurs analytiques commutants, en dimension 3 ou 4: Existence de zéros communs. Bol. Soc. Brasil. Mat. **1992**, 22, 215–247.

23. This paper was inspired by a remarkable result of C. Bonatti, which does not require compactness of M. "The demonstration of this result involves a beautiful and quite difficult local study of the set of zeros of X, as an analyticY-invariant set." — Molino, P. Review of Bonatti [22], Math Reviews 93h:57044; Amer. Math. Soc.: Providence, RI, USA, 1993

24. Hirsch, M. Fixed points of local actions of nilpotent Lie groups on surfaces. arXiv:org/1405.2331. 2013.

25. Hirsch, M. Zero sets of Lie algebras of analytic vector fields on real and complex 2-manifolds. 2015. submitted.

26. Hirsch, M. Zero sets of Lie algebras of analytic vector fields on real and complex 2-manifolds. arXiv.1310.0081v2. 2015.

4

CHAPTER

Some Elementary Aspects of 4-Dimensional Geometry

J. Scott Carter * and David A. Mullens

Department of Mathematics and Statistics, University of South Alabama, Mobile, AL 36688, USA

ABSTRACT

We indicate that Heron's formula (which relates the square of the area of a triangle to a quartic function of its edge lengths) can be interpreted as a scissors congruence in four-dimensional space. In the process of demonstrating this, we examine a number of decompositions of hypercubes, hyper-parallelograms and other elementary four-dimensional solids.

Keywords: Heron's formula; scissors congruence; Nicomachus's theorem; hypercube; four-dimensional geometry; hyper-solids

1. INTRODUCTION

Heron's formula gives the explicit relationship among the area, A, of a triangle and its edge lengths [1]. We assume, now and throughout, that a triangle with edge lengths, a, b, c, where $0 < a \leq b \leq c$, is given. The semi-perimeter is the quantity: $S = (a + b + c)/2$. One representation of the formula is:

$$A = \sqrt{S(S-a)(S-b)(S-c)}.$$

The formula was written in Heron's (Hero's) book Metrica *circa* 60 AD. According to Wikipedia [2], the formula may have been known to Archimedes more than two centuries earlier. However, Heron's book was a compendium of a number of known results. Here, we square both sides of the equation and multiply by the denominator to get the following expression:

$$16A^2 = (a+b+c)(-a+b+c)(a-b+c)(a+b-c).$$

This can be considered an expression that relates the four-dimensional volumes of a number of hyper-solids. As the text develops, we will further manipulate this expression all the while maintaining a four-dimensional awareness of the algebraic manipulations that are performed.

Our primary purpose here is to indicate a four-dimensional scissors congruence among the hyper-solids that are involved in Heron's formula. However, in the process, we will also develop several secondary goals along the way. These are our own interpretations of conversations with several people. Most importantly, we hope to develop the reader's intuition and interest in elementary four-dimensional geometry.

In particular, many algebraic identities, such as the binomial and multinomial theorems, have elegant and useful interpretations in terms of relating the higher dimensional volumes of various n-cubes. The symmetries in the expressions are manifest in the symmetries among the pieces of these decompositions.

We remark here that the algebraic proof of Heron's formula given in this form goes as follows. Suppose that we arrange the triangle in the coordinate plane as indicated in Figure 1.

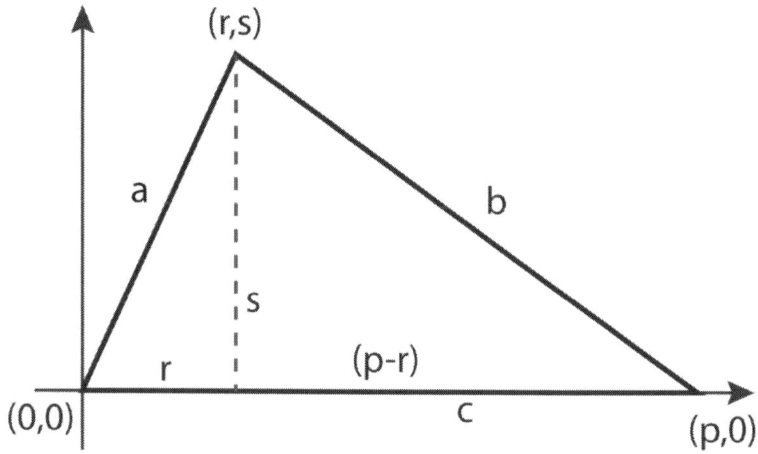

Figure 1. A standard triangle of side lengths $a \leq b \leq c$.

In this way, one vertex is at the origin; the longest side (the side of length c) lies along the x-axis with a vertex at the point $(p, 0)$ (here, a different coordinate is chosen, so that the new variables are of the same flavor). The remaining vertex lies at the point (r, s). In this way, $a^2 = r^2 + s^2$ and $b^2 = s^2 + (p - r)^2$. The left-hand-side of Heron's formula can be shown to be $4p^2s^2$. The right-hand-side reduces to:

$$\text{RHS} = 2a^2b^2 + 2a^2c^2 + 2b^2c^2 - a^4 - b^4 - c^4.$$

We expand the terms on the right-hand-side as follows:

$$\begin{aligned} 2a^2b^2 &= 2(r^2 + s^2)((p - r)^2 + s^2) \\ &= 2r^2(p - r)^2 + 2s^2(p - r)^2 + 2r^2s^2 + 2s^4. \\ 2a^2c^2 &= 2(r^2 + s^2)p^2 \\ &= 2r^2p^2 + 2s^2p^2. \end{aligned}$$

$$
\begin{aligned}
2b^2c^2 &= 2((p-r)^2+s^2)p^2 \\
&= 2(p-r)^2p^2+2s^2p^2. \\
-a^4 &= -(r^2+s^2)(r^2+s^2) \\
&= -r^4-2r^2s^2-s^4. \\
-b^4 &= -((p-r)^2+s^2)((p-r)^2+s^2) \\
&= -(p-r)^4-2s^2(p-r)^2-s^4. \\
-c^4 &= -p^4.
\end{aligned}
$$

We cancel like terms:

$$
\begin{aligned}
RHS &= 2r^2(p-r)^2 \\
&+ 2r^2p^2+2s^2p^2 \\
&+ 2(p-r)^2p^2+2s^2p^2 \\
&+ (-r^4) \\
&+ (-(p-r)^4) \\
&+ (-p^4).
\end{aligned}
$$

After regrouping:

$$\begin{aligned} RHS &= 4s^2p^2 \\ &+ 2r^2(p-r)^2 + 2(p-r)^2p^2 - (p-r)^4 \\ &+ (-p^4) + 2r^2p^2 - r^4. \end{aligned}$$

We factor the last line:

$$\begin{aligned} RHS &= 4s^2p^2 \\ &+ 2r^2(p-r)^2 + 2(p-r)^2p^2 - (p-r)^4 \\ &+ (-(p^2-r^2)^2) \\ &= 4s^2p^2 \\ &+ 2r^2(p-r)^2 + 2(p-r)^2p^2 - (p-r)^4 \\ &+ -(p-r)^2(p+r)^2. \end{aligned}$$

and combine like terms again:

$$\begin{aligned} RHS &= 4s^2p^2 \\ &+ (p-r)^2[2r^2 + 2p^2 - (p-r)^2 - (p+r)^2] \\ &= 4s^2p^2. \end{aligned}$$

As such, the formula has been proven. However, we point out that even in our most quiet moments, the simplification:

$$(a+b+c)(a+b-c)(a-b+c)(-a+b+c) = 2a^2b^2 + a^2c^2 + b^2c^2 - a^4 - b^4 - c^4$$

is quite tedious. After some preliminary work on developing higher dimensional models and the reader's intuition on four-dimensional geometry, our first task will be to show how to use four-dimensional geometry to encode the simplification. The second task that we endeavor upon is to capitalize upon a scissors congruence proof of the Pythagorean theorem, to expand the four-dimensional solids of size a^4, b^4 and a^2b^2. Each of a^2 and b^2 is the square of a hypotenuse of a right triangle and can each be written as a sum of squares. To do so, one decomposes the square figure into five pieces and reassembles them. In four-dimensions, the quartics a^4, b^4 and a^2b^2 are decomposed into twenty-five pieces and reassembled as the sum of four hypercubes (or more precisely, hyper-rectangles). The grouping and factoring that is indicated above can also be seen in terms of four-dimensional scissors congruences. Meanwhile, there is a little work to rearrange the factorization $4s^2p^2$ from four copies of a four-dimensional figure that is a parallelogram times itself into a hyper-rectangle.

In Section 2, we discuss the n-cube from several points of view, including its representation as a configuration space. Then, we decompose it into right n-dimensional simplices. This decomposition gives Nicomachus's theorem [3]. Our interpretation yields that the cube $[0, a]^4$ can be decomposed as the union of four congruent figures, each of the form of a right isosceles triangle times itself. In Section 3, we discuss the multinomial theorem as another decomposition of the n-cube, and we use this idea to organize the computation of the right-hand side of Heron's formula. In Section 4, we demonstrate that the proof of the Pythagorean theorem can be adjusted to a proof of a four-dimensional result: if $z^2 = x^2 + y^2$ and $u^2 + v^2 = w^2$, then the hyper-cube $[0, z] \times [0, z] \times [0, w] \times [0, w]$ can be cut into 25 pieces that can be reassembled into a figure that is the product of a pair of sums of squares. This is the geometric interpretation of the identity:

$$z^2w^2 = x^2u^2 + x^2v^2 + y^2u^2 + y^2v^2.$$

In Section 5, we tie up some loose ends in the algebraic proof above. Section 6 discusses the left-hand-side of the equation, and Section 7 is a conclusion.

2. THE *N*-DIMENSIONAL CUBE

From a mathematician's perspective, the n-dimensional unit cube is a very natural object. It is written as $[0, 1]^n$ and can be considered as the set of functions from $\{1, 2,..., n\}$ to the unit interval. A given function is determined by its set of values, and these are expressed as coordinates $(x_1, x_2,..., x_n)$, where x_j is the value of a given function on $j \in \{1, 2,..., n\}$. For $j = 1, 2,..., n$, each x_j is bounded between zero and one. We occasionally write $0 \leq x_1, x_2,..., x_n \leq 1$ to indicate the collection of such relations. In the rest of this section, we develop a number of metaphors, so that different intuitions can be developed.

2.1. Fingers

Before we consider the unit n-cube in general, let us consider binary digits: your fingers, each in one of two possible states. Hold up your right hand, and make a fist. Now, sequentially start from your thumb and proceed through each of the remaining digits. Hold up your thumb as if hitch-hiking or simply giving a "thumbs-up"; retract it. Now, hold up your index finger, and retract it. Do so with each of your middle, annular and your baby finger (pinky). Hold up some pair of fingers. In how many ways can two fingers be held up (10)? How many ways can you hold up three fingers (again, 10)? How about four fingers (five)?

These are digits since they are fingers. They are binary since they are either up or down. Assign the number $2^0 = 1$ to your thumb; assign $2^1 = 2$ to your index finger; assign $2^2 = 4$ to your middle finger; assign eight to your ring finger; and assign 16 to your pinky. When you consider all possible finger configurations on one hand, you can see that you can represent each integer from zero to 31 as the sum of the values of the upheld fingers in each finger configuration.

In Figure 2, we have arranged the first eight of these configurations at the vertices of a cube in three-dimensional space.

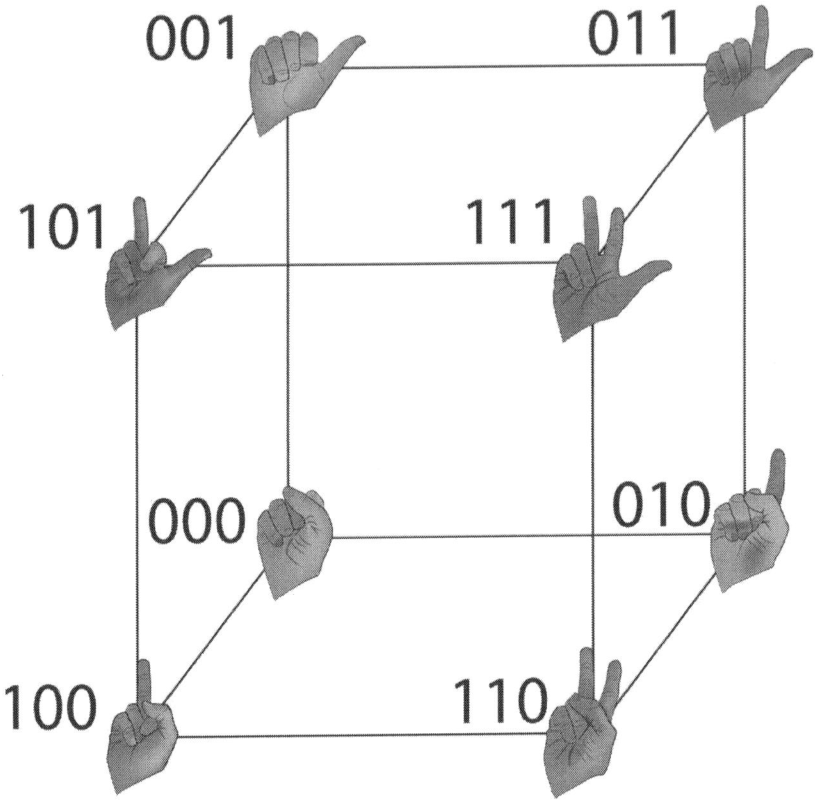

Figure 2. The vertices of a cube correspond to binary digits.

In returning to the description of the unit n-dimensional cube as $[0, 1]^n$—the set of functions from $\{1,..., n\}$ into the unit interval—we observe that the binary sequences are the extreme values of such functions: $x_j = 0$ or $x_j = 1$ for each $j = 1,..., n$.

2.2. Kitchen Drawers

Imagine a small kitchen that has exactly four drawers. Each drawer can be open or closed. As such, we can describe the state of the kitchen cabinet as a binary sequence. However, let us suppose that each drawer can be open any distance from zero cubits up to fully open at a cubit. By convention, a cubit is roughly half a meter.

The units are chosen to appear to be realistic from the point of view of a kitchen. Let us label the drawers with integers one through four. Then, the configurations of the set of drawers corresponds to a function from {1, 2, 3, 4} into [0, 1]. Therefore, the possible states for this kitchen cabinet correspond to the points of a four-dimensional cube. We call this cube the hypercube.

2.3. An Archive

It is easy to generalize from the kitchen cabinet to a large archive in a bureaucracy. This archive has a large number of cabinets, say n of them. Each cabinet consists of a single drawer. The cabinets lie along the south wall of a building and are numbered east-to-west from one through n. The drawers can open up to a full meter. Therefore, the set of possible configurations of the cabinets corresponds to the unit n-dimensional cube $[0, 1]^n$, since if x_j is the distance (in meters) that the j-th drawer is open, then $0 \leq x_j \leq 1$. The configuration of any one drawer is independent of the configuration of all other drawers.

We begin this subsection with a simple case. Consider the possible configurations of a pair of drawers. Let x denote the distance that the first drawer is opened, and let y denote the distance that the second drawer is opened. Of course, $0 \leq x, y \leq 1$, that is each drawer can be opened any amount from zero meters to one meter. This situation can be subdivided into two cases. Either the drawer on the left is no more open than the drawer on the right is, or *vice versa*. Thus, either $0 \leq x \leq y \leq 1$ or $0 \leq y \leq x \leq 1$. These cases are not disjoint; their coincidence occurs when both drawers are equally open: $x = y$. The set of all configurations of the drawers corresponds to a unit square. We have decomposed the square as the union of two isosceles right triangles whose legs are a unit length. They intersect along their diagonals.

For example, the point (1/4, 1/2) in the unit square corresponds to the configuration in which the drawer on the left is open half as much as the drawer on the right, which is half-open. See Figure 3.

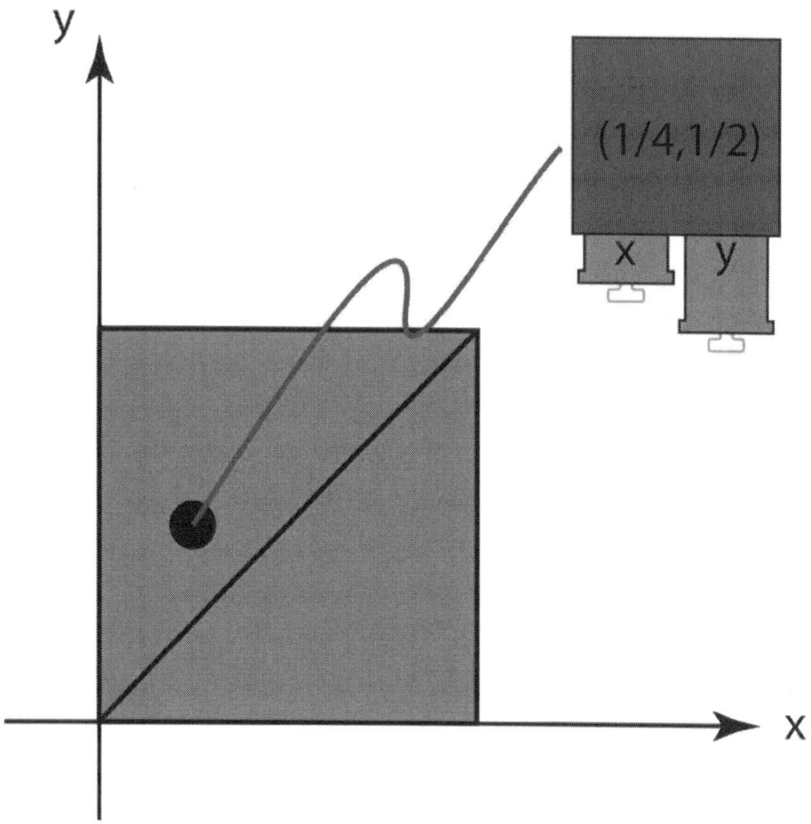

Figure 3. The square as a pair of open drawers.

2.4. Triangles, Tetrahedra, and So Forth

Now, consider three drawers. Let x, y and z denote the amount that each is pulled out. As before, $0 \le x, y, z \le 1$, but there is no relation among the openness of the drawers in general. We decompose this situation into six configurations:

$$0 \le x \le y \le z \le 1, \quad 0 \le y \le x \le z \le 1, \quad 0 \le x \le z \le y \le 1,$$

$$0 \le z \le x \le y \le 1, \quad 0 \le y \le z \le x \le 1, \quad \& \quad 0 \le z \le y \le x \le 1.$$

as indicated in Figure 4. Using the coordinates of the three-dimensional cube that are indicated, the cube has been decomposed into six congruent tetrahedra. Each tetrahedron has three edges with length one, two edges of length $2\sqrt{2}$ and one edge of length $3\sqrt{3}$. All of the tetrahedra intersect along the edge of length $3\sqrt{3}$, since this line corresponds to the configuration of drawers in which all are equally open. The interior triangular faces with edge lengths one, $2\sqrt{2}$ and $3\sqrt{3}$correspond to configurations of drawers in which at least two are equally open and the other drawer is either open no more or no less than the two equal drawers. The exterior triangular faces that have edge lengths 1, 1 and $2\sqrt{2}$ correspond to exactly one drawer being fully open or fully closed.

In full generality, we have n drawers with drawer number j open to a distance x_j. In this case, $0 \le x_1, x_2, ..., x_n \le 1$. There are $n! = n(n-1)!$ (where $0! = 1$) possible orderings of the distances that the drawers can be open. To see this, observe that the$j$-th drawer may be pulled out a greater distance than all of the others. Thus, there are n possible ways for one drawer to be more extended than all of the rest. By induction, the remaining $(n - 1)$ drawers are ordered in $(n - 1)!$ configurations.

A right n-dimensional simplex consists of the point set:

$$\Delta^n = \{(x_1, x_2, \ldots, x_n) \in [0,1]^n : 0 \le x_1 \le x_2 \le \cdots \le x_n \le 1\}.$$

The n-dimensional cube can be decomposed into $n!$ such simplices that are all congruent. Therefore, the n-dimensional volume of such a simplex is 1n!1n!.

There is a typographical trick called the $L7$-trick (*cf.* the second verse to *Woolly Bully* as sung by Sam the Sham and the Pharaohs), to obtain the set of vertices for each simplex. Let n be fixed. Consider a row of n 0s (separated, if you like, by commas): 00...0. Furthermore, construct a column of n 1s. Now, the column of 1s can be appended below any one of the 0s. Thus, there are n possible ways of doing so. For each such shape, we can insert the $(n - 1)!$ shapes that exist by induction. The first step of the induction is to consider the array [01][01]. In the next step, we have the arrays:

$$\begin{bmatrix} 0 & 0 \\ & 1 \\ & 1 \end{bmatrix} \quad \begin{bmatrix} 0 & 0 \\ 1 & \\ 1 & \end{bmatrix}$$

which can be filled as:

$$\begin{bmatrix} 0 & 0 \\ 0 & 1 \\ 1 & 1 \end{bmatrix} \quad \begin{bmatrix} 0 & 0 \\ 1 & 0 \\ 1 & 1 \end{bmatrix}$$

The rows are the coordinates of the vertices of the two triangles $0 \leq x \leq y \leq 1$ and $0 \leq y \leq x \leq 1$ into which the unit square was decomposed. For the cube, we start from the three arrays:

$$\begin{bmatrix} 0 & 0 & 0 \\ & & 1 \\ & & 1 \\ & & 1 \end{bmatrix} \quad \begin{bmatrix} 0 & 0 & 0 \\ & 1 & \\ & 1 & \\ & 1 & \end{bmatrix} \quad \begin{bmatrix} 0 & 0 & 0 \\ 1 & & \\ 1 & & \\ 1 & & \end{bmatrix}$$

and fill them respectively with the two arrays above.

$$\begin{bmatrix} 0 & 0 & 0 \\ 0 & 0 & 1 \\ 0 & 1 & 1 \\ 1 & 1 & 1 \end{bmatrix} \begin{bmatrix} 0 & 0 & 0 \\ 0 & 1 & 0 \\ 0 & 1 & 1 \\ 1 & 1 & 1 \end{bmatrix} \begin{bmatrix} 0 & 0 & 0 \\ 1 & 0 & 0 \\ 1 & 0 & 1 \\ 1 & 1 & 1 \end{bmatrix}$$

$$\begin{bmatrix} 0 & 0 & 0 \\ 0 & 0 & 1 \\ 1 & 0 & 1 \\ 1 & 1 & 1 \end{bmatrix} \begin{bmatrix} 0 & 0 & 0 \\ 0 & 1 & 0 \\ 1 & 1 & 0 \\ 1 & 1 & 1 \end{bmatrix} \begin{bmatrix} 0 & 0 & 0 \\ 1 & 0 & 0 \\ 1 & 1 & 0 \\ 1 & 1 & 1 \end{bmatrix}$$

In the next step, each of these six matrices are fed into the four patterns:

$$\begin{bmatrix} 0 & 0 & 0 & 0 \\ & & & 1 \\ & & & 1 \\ & & & 1 \end{bmatrix} \begin{bmatrix} 0 & 0 & 0 & 0 \\ & & 1 & \\ & & 1 & \\ & & 1 & \end{bmatrix} \begin{bmatrix} 0 & 0 & 0 & 0 \\ & 1 & & \\ & 1 & & \\ & 1 & & \end{bmatrix} \begin{bmatrix} 0 & 0 & 0 & 0 \\ 1 & & & \\ 1 & & & \\ 1 & & & \end{bmatrix}$$

The resulting rows of each matrix are the five vertices of each four-simplex. The union of these fills the four-cube. The situations are illustrated in Figure 5. In this figure, six copies of each three-dimensional cube $x = 1, y = 1, z = 1$ and $w = 1$ are illustrated. In each such cube, a tetrahedron is drawn, and this is to be coned to the origin in the figure. The rows of the matrices that appear to the left of each hypercube indicate the coordinates of the vertices of the four-simplices, which are drawn as small purple spheres.

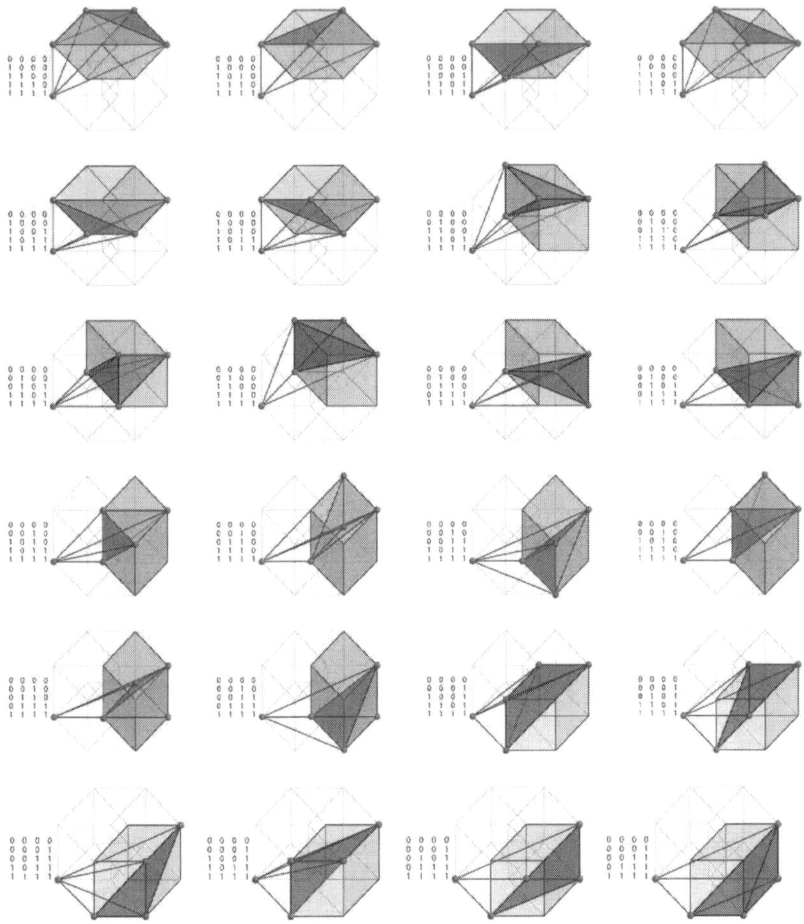

Figure 5. The twenty-four four-simplices that fill the four-cube.

2.5. Pyramidal Sets

In the unit n-cube, $[0, 1]^n$ consider the set:

$$P_n = \{(x_1, \ldots, x_n) : 0 \leq x_j \leq 1 \text{ (for all } j = 1, \ldots, n) \;\&\; x_i \leq x_n \leq 1 \text{ (for all } i = 1, \ldots, n-1)\}.$$

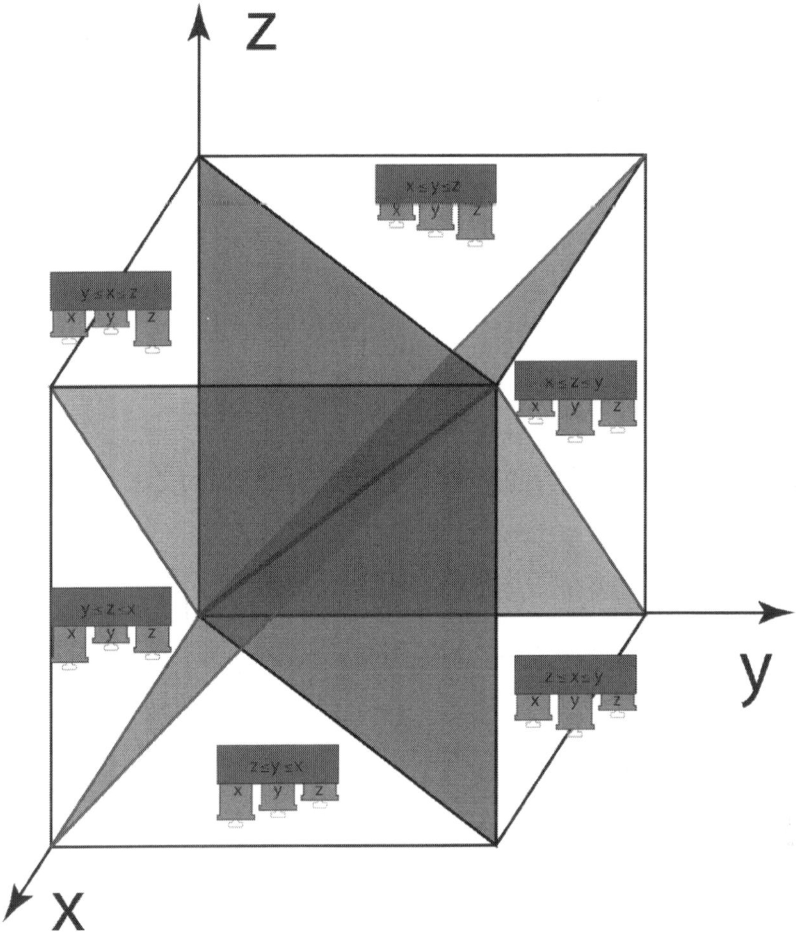

Figure 4. Decomposing the three-cube as the union of six tetrahedra.

For any fixed value $x_n = C$, the cross-sectional set $\{(x_1,..., x_{n-1}, C) : 0 \leq x_j \leq C\}$ is an $(n - 1)$-dimensional cube $[0, C]^{n-1}$ whose volume (we will use the term volume to mean n-dimensional volume) then is C^{n-1}.

Considering the situation of drawers, we can imagine a "high priority drawer", which can be opened any distance between zero and one meters. If it is opened to C meters, then the remaining drawers can only be opened that far. Alternatively, we can imagine an alternative situation in which a king or queen (sovereign) is seated in one of n-chairs. All of the subjects are also seated, and no one is to stand taller than the sovereign. As the sovereign stands, all of the inferiors attempt to stand. The set of configurations of partially standing people corresponds to the configurations of partially opened drawers and of the points in the pyramidal set P_n. Here, note that the subscript n on the expression P_n indicates that the last coordinate is the position of the sovereign.

More generally, let:

$$P_k = \{(x_1, \ldots, x_n) : 0 \leq x_j \leq 1 \text{ (for all } j = 1, \ldots, n) \;\&\; x_i \leq x_k \leq 1 \text{ (for all } i \neq k)\}.$$

Observe that the n-cube is decomposed into n congruent pyramids. One can then interpret this decomposition as a geometric manifestation of the fact that:

$$\int_0^1 x^{n-1}\, dx = \frac{1}{n}.$$

This interpretation was given in [4] and earlier in [5].

2.6. The Cone on a Cube Is a Square of a Triangle

A favorite formula from discrete mathematics is Nicomachus's theorem:

$$\sum_{k=1}^{n} k^3 = \left(\binom{n+1}{2}\right)^2.$$

One usually proves this by induction. Since:

$$n^2(n+1)^2/4 + (n+1)^3 = (n+1)^2\left[n^2/4 + (n+1)\right] = \frac{(n+1)^2}{4}(n^2 + 4n + 4),$$

the proof follows easily. That formula is often used to compute the integral ∫10x3∫01x3 dx as the limit $lim_{n\to\infty}\left(\sum_{k=0}^{N}k^3\right)\frac{1}{N^3}$.
The last expression comes about by subdividing the unit interval into N equal subintervals and approximating the area under the curve $y = x^3$ by rectangles of base $1/N$ and height (k/N^3).

Dylan Thurston pointed out to the first named author that one could consider the formula continuously. Thus, the four-dimensional figure that is a cone on a cube has the same four-dimensional volume as the square of a triangle. Dylan's argument was developed over a dinner that he had with Dave Bayer and Walter Neumann. This section presents our modification of another dinner-time description that was given by Dror Bar-Natan in Hanoi on some evening in the period of 6–12 August 2007.

First, we consider the product of a right isosceles triangle with itself. This will be:

$$\{(x, y, z, w) : 0 \le x \le y \le 1 \,\&\, 0 \le z \le w \le 1\}.$$

As a configuration of drawers, the first drawer is no more open than the second, while the third drawer is no more open than the fourth. Alternatively, we consider a pair of married couples in which neither husband speaks more than his own wife, but either husband may speak more than the wife of the other.

Now, the hypercube $[0, 1]^4$ can be decomposed into four congruent figures:

$$\Delta_{x\le y} \times \Delta_{z\le w} = \{(x, y, z, w) : 0 \le x \le y \le 1 \,\&\, 0 \le z \le w \le 1\}.$$

$$\Delta_{y\le x} \times \Delta_{z\le w} = \{(x, y, z, w) : 0 \le y \le x \le 1 \,\&\, 0 \le z \le w \le 1\}.$$

$$\Delta_{x\le y} \times \Delta_{w\le z} = \{(x, y, z, w) : 0 \le x \le y \le 1 \,\&\, 0 \le w \le z \le 1\}.$$

and

$$\Delta_{y\le x} \times \Delta_{w\le z} = \{(x, y, z, w) : 0 \le y \le x \le 1 \,\&\, 0 \le w \le z \le 1\}.$$

See Figure 6 for an illustration.

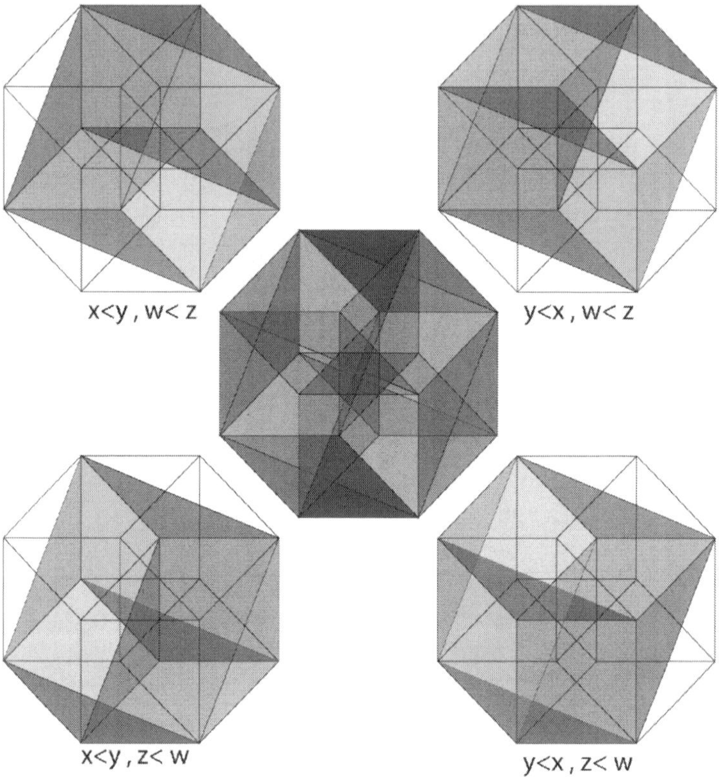

Figure 6. Decomposing the four-cube as the union of four copies of $\Delta \times \Delta$.

Additionally, it also can be decomposed into the four pyramidal sets:

$$P_4 = \{(x, y, z, w) : \ 0 \leq x, y, z \leq w \leq 1\},$$

$$P_3 = \{(x, y, z, w) : \ 0 \leq x, y, w \leq z \leq 1\},$$

$$P_2 = \{(x, y, z, w) : \ 0 \leq x, z, w \leq y \leq 1\},$$

and

$$P_1 = \{(x, y, z, w) : \; 0 \leq y, z, w \leq x \leq 1\}.$$

Now, we decompose, $\Delta_{x \leq y} \times \Delta_{z \leq w}$ into six four-simplices in which:

$$0 \leq x \leq y \leq z \leq w \leq 1, \quad 0 \leq x \leq z \leq w \leq y \leq 1,$$

$$0 \leq x \leq z \leq y \leq w \leq 1, \quad 0 \leq z \leq x \leq w \leq y \leq 1,$$

$$0 \leq z \leq x \leq y \leq w \leq 1, \quad 0 \leq z \leq w \leq x \leq y \leq 1.$$

and we decompose P_4 into six four-simplices:

$$0 \leq x \leq y \leq z \leq w \leq 1, \quad 0 \leq y \leq x \leq z \leq w \leq 1,$$

$$0 \leq x \leq z \leq y \leq w \leq 1, \quad 0 \leq y \leq z \leq x \leq w \leq 1,$$

$$0 \leq z \leq x \leq y \leq w \leq 1, \quad 0 \leq z \leq y \leq x \leq w \leq 1;$$

the three of which listed to the left correspond to the three that are listed to the left above. Therefore, half of the volume of the pyramid P_4 coincides with half of the volume of $\Delta_{x \leq y} \times \Delta_{z \leq w}$. The other half of P_4 coincides with half the volume of $\Delta_{x \leq y} \times \Delta_{z \leq w}$. Dylan describes this as cutting the cube $\{(x, y, z) : 0 \leq x, y, z \leq 1\}$ into two pyramids (($x \leq y$) and ($y \leq x$)), considering the cone on each, twisting one and reassembling it to the square of the triangle.

Observe that a four-cube of the form $[0, a]^4$ also can be cut into the union of four figures each of which is the product of two isosceles right triangles of edge length a, a and 2√a2a. This decomposition is considered as a special case of Heron's formula, for in this case $a = b$ and c = a2√a2. The right-hand side reduces to $4a^4$.

Therefore, sixteen copies of a right isosceles triangle times itself are reassembled into four hyper-cubes.

3. THE DISTRIBUTIVE LAW, THE MULTINOMIAL THEOREM AND SCISSORS CONGRUENCES

Two n-dimensional polyhedra are said to be scissors congruent if and only if they each can be cut into a union of congruent polyhedra. For the example above, we demonstrated that the n-cube is scissors congruent to a union of right n-simplices, and we demonstrated a scissors congruence between the product of a right isosceles triangle and itself with the cone on a cube. In this paper, we are primarily concerned with four-dimensional figures that are hypercubic, cones on three-dimensional polyhedra and the two-fold product of polygons.

Famously, Max Dehn [6,7] demonstrated that a regular tetrahedron of unit volume is not scissors congruent with a unit cube.

To begin this section, in which an n-dimensional cube with variable edge lengths is decomposed into several smaller sub-cubes, we first consider the case $n = 1$.

A line segment whose length is the sum $x + y$ of two positive quantities x and y can be cut into two segments with the segment of length x on the left and the segment of length y on the right (or *vice versa*). Similarly, suppose that $x < y$ and a segment of length $y - x$ is given. It could be produced by means of cutting a segment of length x from the longer segment. Of course, we can, in general, discuss directed segments with the direction "right" indicating positive length quantities and the direction "left" used to indicate negative quantities. In this way, each of $x + y$, $x - y$ and $y - x$ can be thought of as being built by composing directed segments of lengths x and y. It is conceptually easier to think in this fashion rather than explicitly removing segments and hypercubes whose total volumes in algebraic expressions happen to be negative.

In the case that a given rectangular area of dimensions $(x + y)$-by-z can be decomposed as a pair of rectangular areas x-by-$z$and y-by-z, as above, we can indicate $(x - y)z$ as the difference between areas xz and yz. We further exploit the distributive law as a scissors congruence when considering the case $(x + y)$-by-$(z + w)$. This rectangular area is decomposed as the union of four rectangles of size xz, xw, yz and yw. Some of these scissors congruences are indicated in Figure 7. We leave the reader to imagine others.

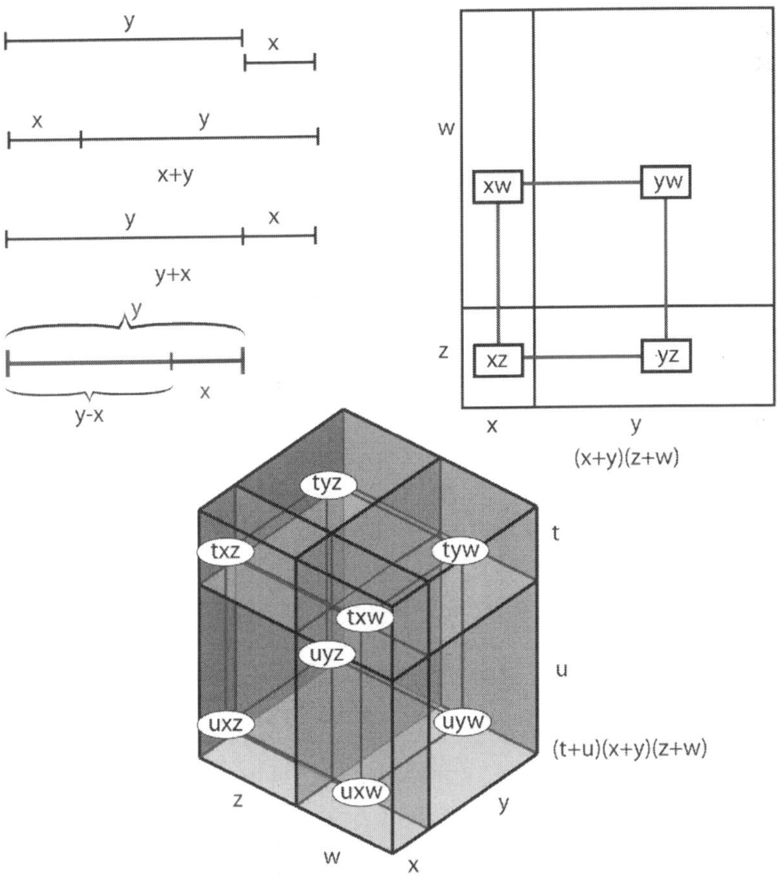

Figure 7. The distributive law as a scissors congruence.

For convenience within cubes, hypercubes and beyond, we dualize the pictures by indicating a vertex with a label that indicates its volume (and is drawn red if the net volume is negative) and connecting these by an edge if they share a codimension one face. If they share a codimension two face, then the dual vertices will bound a polygon (always a rectangle in our cases), and so forth. The language of shared polygonal vertices, edges, faces, and so forth, is initially quite awkward (thus, the term "codimension" is used), but in our case, the situation is a bit more straight-forward than one might presume.

Our main interest here is to consider the n-dimensional cube that is decomposed as a product of edges in the form:

$$RH = (a+b+c)(a+b-c)(a-b+c)(-a+b+c).$$

This expression RH is called the right-hand side of Heron's formula. When we use the distributive law to expand the product as the union of 81 hypercubes (more precisely, hyper-rectangles), we want to identify which terms will cancel. It is not terribly difficult to determine that the product reduces to:

$$2a^2b^2 + 2a^2c^2 + 2b^2c^2 - a^4 - b^4 - c^4.$$

"Not terribly difficult" is a relative phrase. An electronic computer can do so easily; a human has to develop some technique to organize and regroup the 81 terms that are initially involved in the expansion. Our purpose here will be to find within the hyper-solid the location of each piece. We will observe that the pieces that cancel can be found on specific affine three-dimensional planes.

To help determine the addresses of the vertices in the dual complex, we will examine the terms in the multinomial expansion:

$$(x_1 + x_2 + \cdots + x_k)^n.$$

This situation is more general in that we are considering an n-dimensional quantity, but it is too specific for the case at hand, since the terms in the expansion of *RH* involve differing edge lengths. Still, once the addressing problem is solved for the multinomial expansion, it will be easy to apply here.

The quantity $(x_1 + x_2 + \cdots + x_k)^n$ n corresponds to decomposing $[0, \sum_k x_k]^n$ into k^n pieces. The volumes of each piece are a product of x_is, where each x_i is taken from one factor of the decomposition of the edges of the n-cube. Thus, there is a specific n-cube (The correct terminology might be n-rectangle, since the edge lengths vary. Here, we will use the more concise terminology, but elsewhere, we may speak of "hyper-rectangles".) of volume $x_{i_1} x_{i_2} \cdot \cdots \cdot x_{i_n}$, where each $i_\ell \in \{1, \ldots, k\}$, and this ℓ-th factor xiℓxiℓ was selected from the ℓ-th factor in the product expansion:

$$(x_1 + \cdots + x_k) \cdots \underbrace{(x_1 + \cdots + \boxed{x_{i_\ell}} + \cdots + x_k)}_{\ell\text{th factor}} \cdots (x_1 + \cdots + x_k)$$

In order to keep track of these terms, we consider the coordinates of the points in the integral lattice that lie within the cube $[0, (k - 1)]^n$. These coordinates range between (0, 0,..., 0) and $(k - 1, k - 1, \ldots, k - 1)$. We then label the point $(i_1 - 1, i_2 - 1, \ldots, i_n - 1)$ in the integral lattice with the term (xi1xi2·····xin)(xi1xi2·····xin) in the expansion above.

The next step in understanding the multinomial theorem is to count all of the terms that have the same volume (xi1xi2·····xin)(xi1xi2·····xin) as the cube [0,xi1]×[0,xi2]×···×[0,xin]×···×[0,xin] does. Therefore, we can think of {{i1−1,i2−1,...,in−1}} as a multi-set, that is a set possibly with repetitions. The products xi1xi2·····xin and xj1xj2·····xjnxi1xi2·····xinandxj1xj2·····xjn have the same volumes if and only if the corresponding multisets are the same or the sequences $(i_1 - 1, i_2 - 2, \ldots, i_n - 1)$ and $(j_1 - 1, j_2 - 1, \ldots, j_n - 1)$ are permutations of each other. If so, then these points lie along the same $(n - 1)$-dimensional affine plane ∑nj=1zj=K∑j=1nzj=K, where K=(∑nℓ=1xiℓ)−nK=(∑ℓ=1nxiℓ)−n. The value K ranges between zero and $n(k - 1)$.

This last condition is necessary, but not sufficient. For example, in $\mathbb{R}^3$, the plane $x + y + z = 2$ intersects the lattice in the points (2, 0, 0), (0, 2, 0), (0, 0, 2) (which correspond respectively to the products $x_3x_1x_1$, $x_1x_3x_1$ and $x_1x_1x_3$) and the points (1, 1, 0) (1, 0, 1), (0, 1, 1) (corresponding respectively to the products $x_2x_2x_1$, $x_2x_1x_2$ and $x_1x_2x_2$).

In the illustration of Figure 8, the 27 terms in the expansion of (x1+x2+x3)3(x1+x2+x3)3 are grouped according to the intersection with the plane $x + y + z = K$, as also indicated in the Table 1 below.

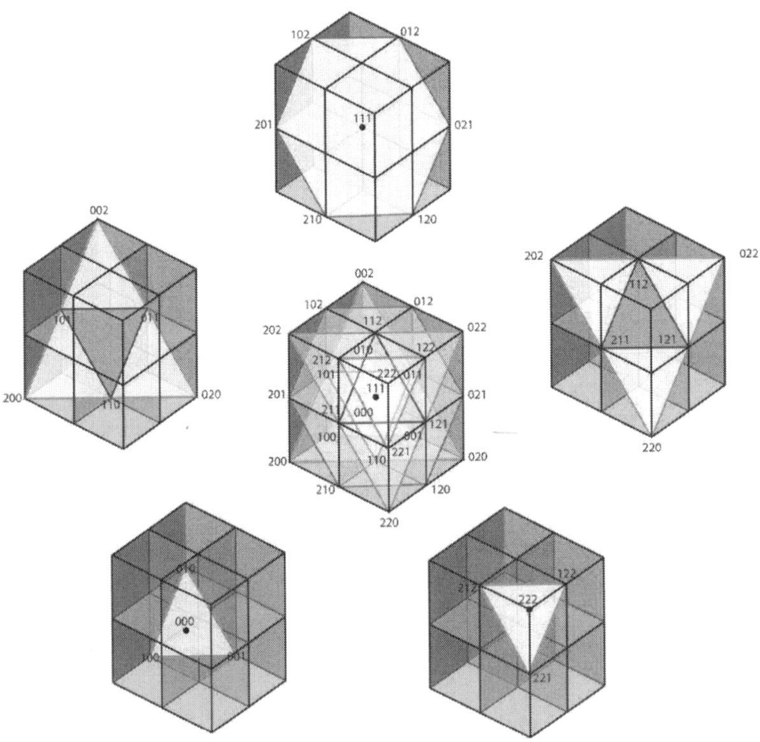

Figure 8. The trinomial theorem $n = 3$. Note that ijk corresponds to $x_{i+1}x_{j+1}x_{k+1}$.

Table 1. The correspondence between coordinates and monomials.

Coordinate	Monomial
$(0,0,0)$	x_1^3
$(1,0,0),(0,1,0),(0,0,1)$	$3x_1^2x_2$
$(2,0,0),(0,2,0),(0,0,2)$	$3x_1^2x_3$
$(1,1,0),(1,0,1),(0,1,1)$	$3x_1x_2^2$
$(1,1,1)$	x_2^3
$(2,2,0),(2,0,2),(0,2,2)$	$3x_1x_3^2$
$(1,1,2),(1,2,1),(2,1,1)$	$3x_2^2x_3$
$(2,2,1),(2,1,2),(1,2,2)$	$3x_2x_3^2$
$(2,2,2)$	x_3^3

The full statement of the multinomial theorem is:

$$(x_1+x_2+\cdots+x_k)^n = \sum_{i_1+i_2+\cdots+i_\ell=n} \left(\frac{n!}{i_1!i_2!\cdots i_\ell!}\right) x_1^{i_1}\cdots x_2^{i_2}\cdots x_\ell^{i_\ell}.$$

We envision the multinomial coefficient as an expression of the symmetry of the intersection of the affine $(n-1)$-plane $\sum_{j=1}^{n} z_j = K$ or, equivalently, the symmetry of the multi-set:

$$\{\{\underbrace{x_1,\cdots,x_1}_{i_1},\underbrace{x_2,\cdots,x_2}_{i_2},\cdots\underbrace{x_\ell,\cdots,x_\ell}_{i_\ell}\}\}.$$

In the approach above, we kept track of each term in the expansion as a product: the first factor in the product that represents a term was taken as a term from the first factor of the product. You may reread that sentence several times if you like, but it is better now simply to consider the product:

$$RH = (a+b+c)(a+b-c)(a-b+c)(-a+b+c).$$

In each term of the expansion, the first factor is positive, the second factor is negative if it is a c, and so forth. We will arrange the individual terms along the affine three-planes $x + y + z + w = K$ for $K = 0, 1, \ldots, 8$ and color the terms blue if they are negative and red if they are positive. Figures 9–16 contain the illustrations. In this way, we need only consider up to 12 terms at any one time, and only among those terms will cancellations occur.

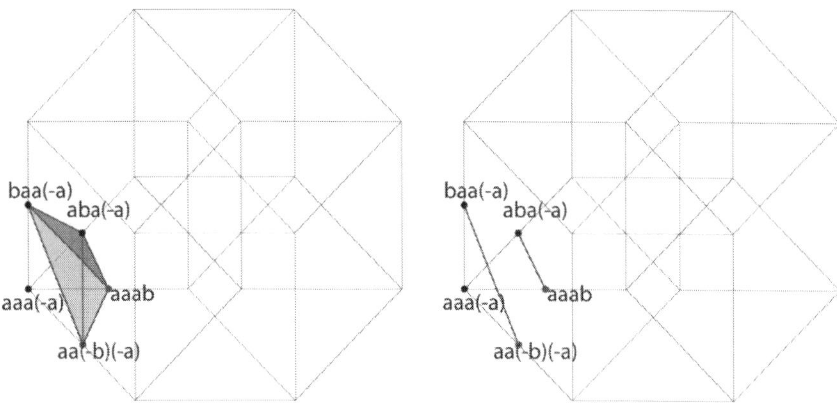

Figure 9. The terms of the form a^4 and a^3b in expansion of the RHS.

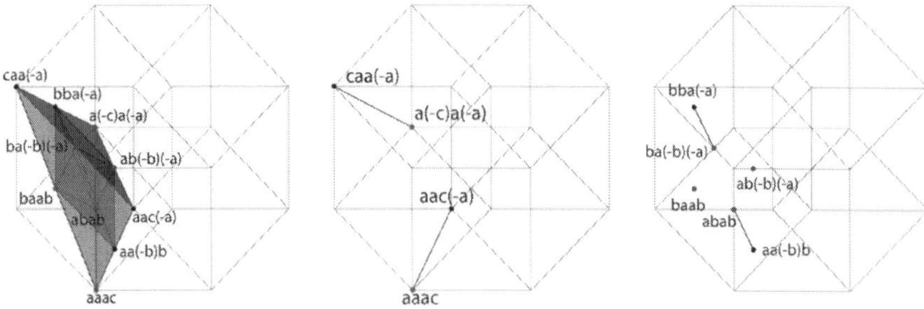

Figure 10. The terms of the form a^3c and a^2b^2 in the expansion of the RHS.

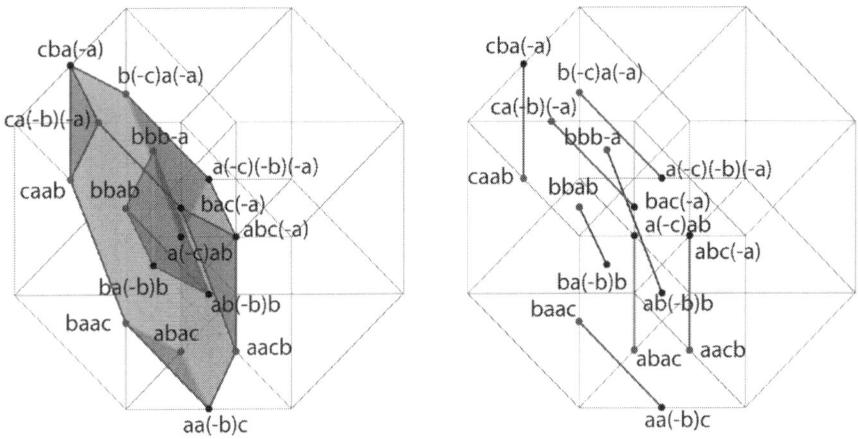

Figure 11. The terms of the form a^2bc and ab^3 in the expansion of the RHS.

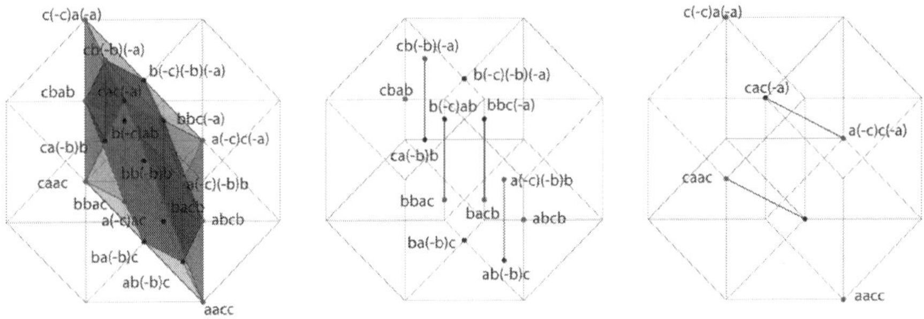

Figure 12. The terms of the form ab^2c and a^2c^2 in the expansion of the RHS.

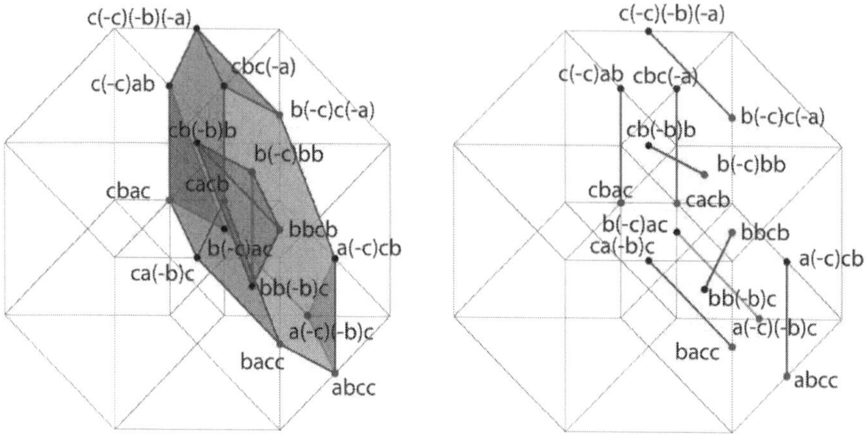

Figure 13. The terms of the form abc^2 and b^3c in the expansion of the RHS.

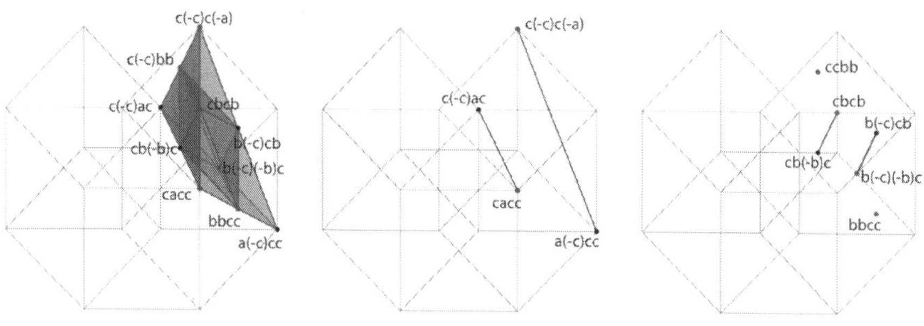

Figure 14. The terms of the form ac^3 and b^2c^2 in the expansion of the RHS.

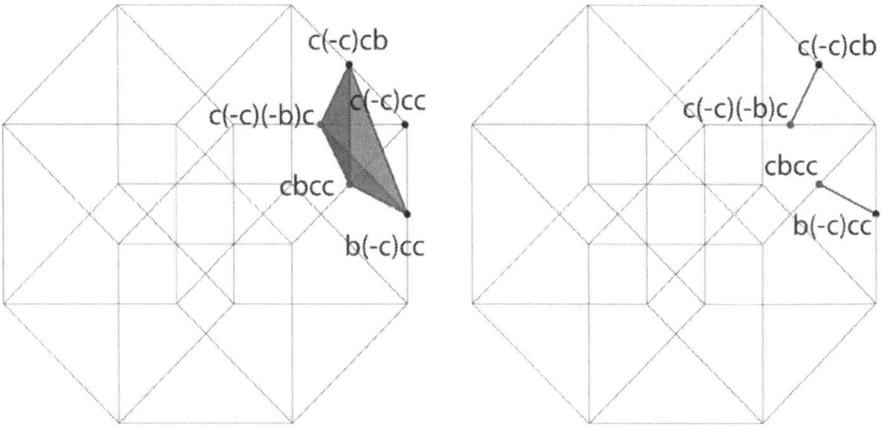

Figure 15. The terms of the form bc^3 and c^4 in the expansion of the RHS.

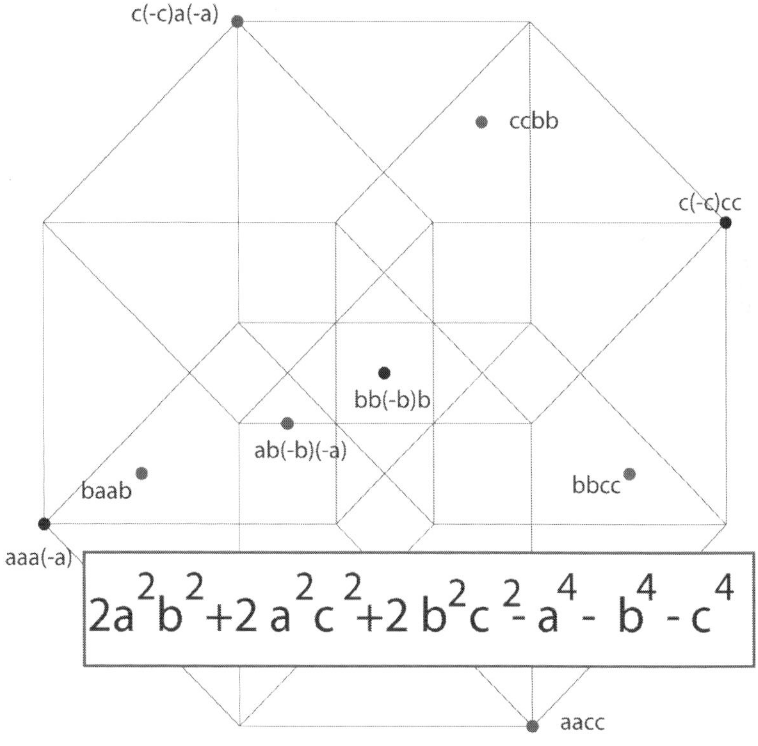

Figure 16. The terms that remain in the expansion of the RHS.

4. THE PYTHAGOREAN THEOREM

The illustration in Figure 17 schematizes a proof of the Pythagorean theorem. Four right triangles with legs of length $x$and y and hypotenuse z are arranged to nearly fill a square whose edge length is z. The missing piece is a square of size $(y - x) \times (y - x)$, where, by convention, $x \leq y \leq z$ (in the case for which $x = y$, the central square is empty). Let us call the copies of the triangles *NE*, *SE*, *SW* and *NW*, depending on their positions with respect to the cardinal directions. The *NE* and *SW*triangles are reassembled to form a rectangle of size $x \times y$; similarly the *NW* and *SE* triangles form a congruent rectangle. In the reassembly, the triangles are only moved by translation. Then, the central square is appended. The union on the central square and the *NW* to the *SE* rectangle together with a portion of remaining rectangle form a square of size $y \times y$. The remaining piece of the other rectangle is a square of size $x \times x$. Thus, the area $z \times z$ can be cut into two squares of size $x \times x$and $y \times y$.

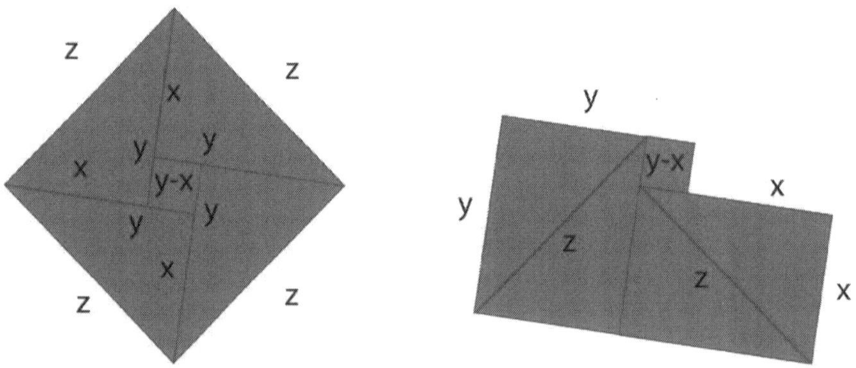

Figure 17. The proof of the Pythagorean theorem.

Now, consider a pair of right triangles. The first has legs of length x and y with $x \leq y$. The second has legs of length u and $v$with $u \leq v$. The length of the hypotenuses are z and w, respectively. We will now consider a hyper-rectangle $R_{zzww} = [0, z] \times [0, z] \times [0, w] \times [0, w]$, and decompose both the $(z \times z)$ and $(w \times w)$ squares into five pieces (four triangles and one small square) as in the proof of the Pythagorean theorem. The hyper-rectangle R_{zzww} is cut into 25 pieces. Two views of this object are illustrated in Figure 18. We encourage the reader to attempt a cross-eyed stereopsis to imagine some of the depth in this illustration.

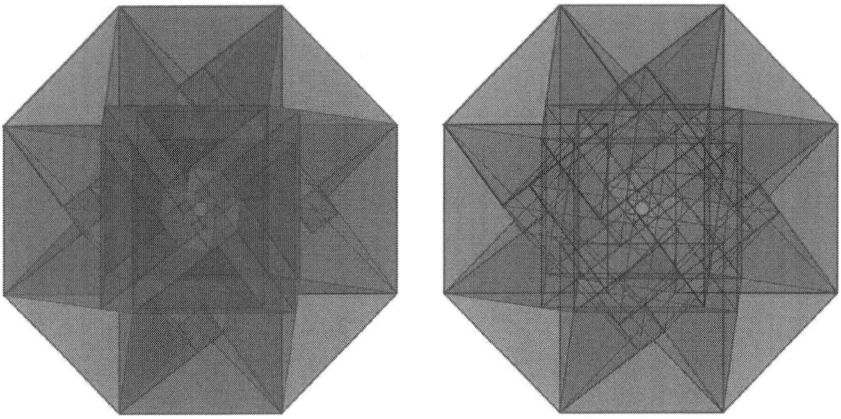

Figure 18. Two views of the hyper-rectangle R_{zzww}.

In our illustrations, we project the *zz* directions into the horizontal and vertical directions of the plane defined by the paper (computer screen) and project the *ww* directions into directions at a 45° angle. In this way, we will speak of the *N*, *S*, *E* and *W* triangles and the *NW*, *NE*, *SW* and *SE* triangles. Furthermore, we will consider the cardinal directions first and the off-cardinal directions second. Then, in Figures 19–23, we illustrate the 25 pieces that will result as planar projections of solids of the form (triangle × triangle), (triangle × square), (square × triangle) and (square × square).

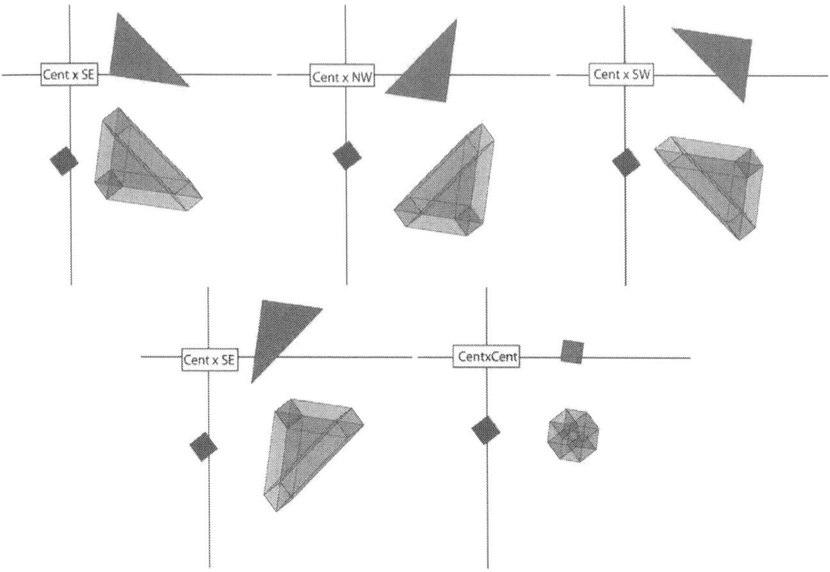

Figure 19. The central pieces of the decomposition of the hyper-rectangle.

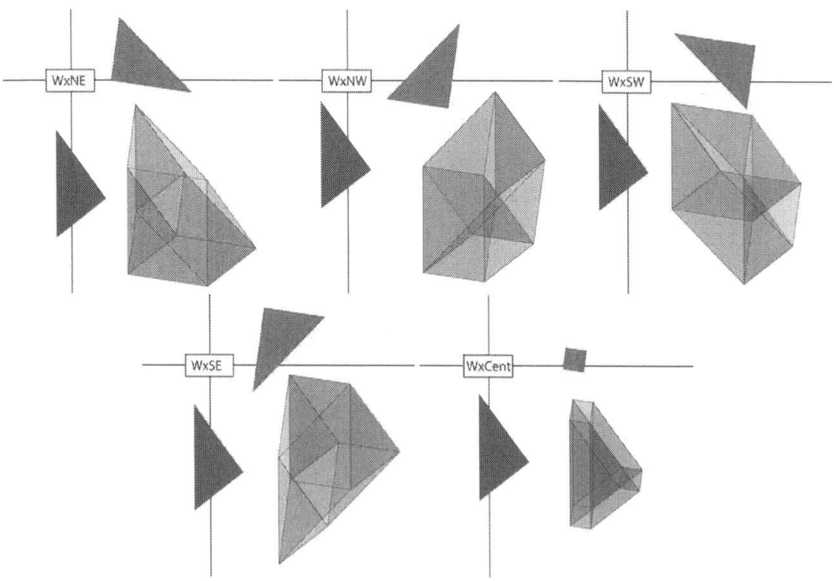

Figure 20. The western pieces of the decomposition of the hyper-rectangle.

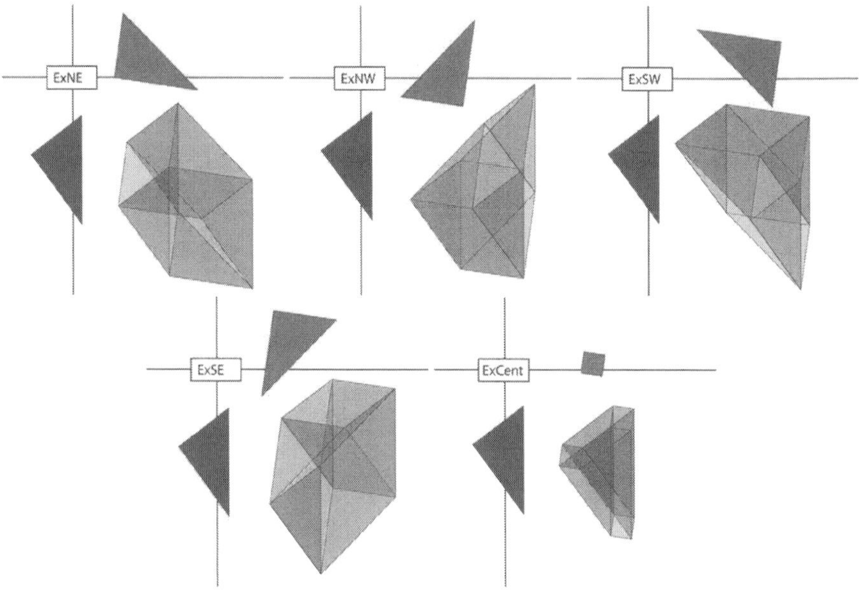

Figure 21. The eastern pieces of the decomposition of the hyper-rectangle.

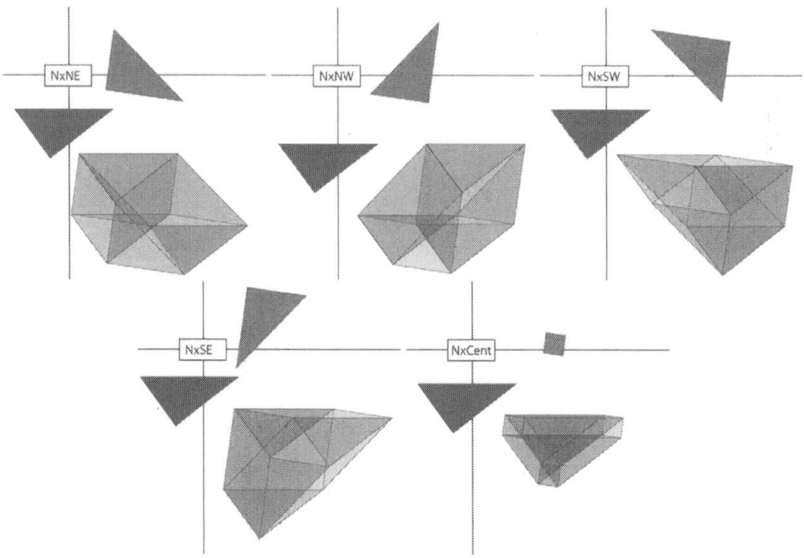

Figure 22. The northern pieces of the decomposition of the hyper-rectangle.

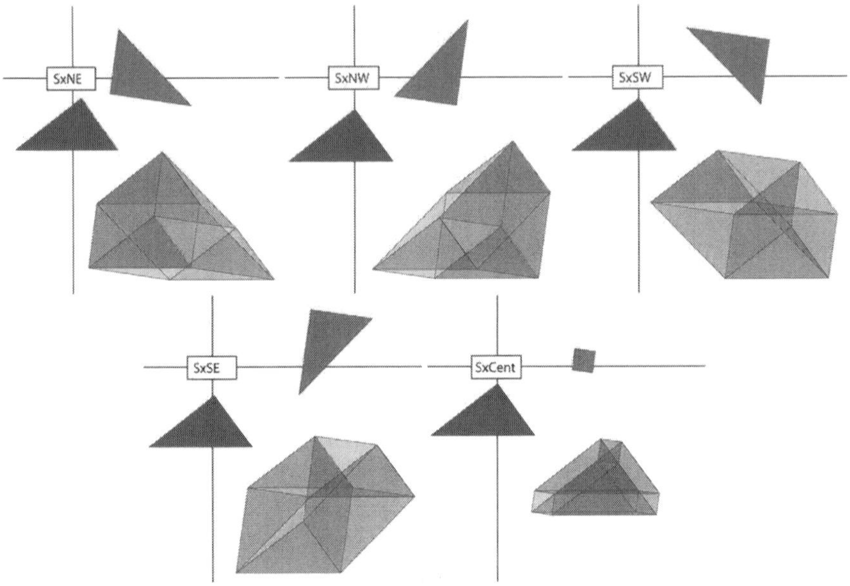

Figure 23. The southern pieces of the decomposition of the hyper-rectangle.

Next, we reassemble these pieces following the reassembly of the Pythagorean theorem. Some of these reassemblies are illustrated in Figures 20–23. In Figure 24, the green northeast/southwest and northwest/southeast triangles times east and west brown triangles are reassembled into two (green) rectangles times (brown) triangles. A similar reassembly occurs inFigure 25. Figure 26 indicates the product of a small green square with each of the (brown) north/south and east/west rectangles and the small brown square. There is a similar figure (not drawn) that is the product of the small brown square and each of the two green rectangles. The small green square times small brown square is only used once. Figure 27indicates how triangles times rectangles are reassembled to be rectangles times rectangles. Then, in Figure 28, the small central hypercube is placed in the one-skeleton of the full reassembly. Figure 29 demonstrates the outlines of the four hypercubes fit into this conglomeration. Finally, Figure 30 shows all four of these square-by-squares inside this conglomeration.

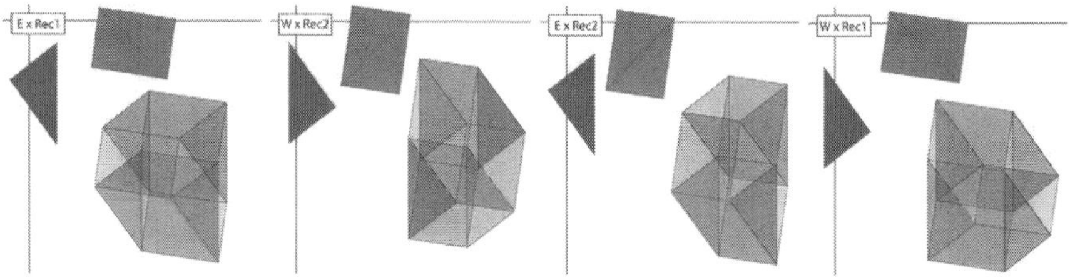

Figure 24. Reassembling the eastern and western triangles times rectangles.

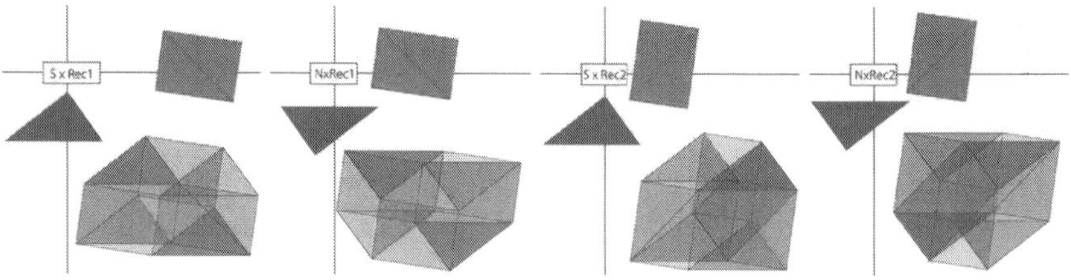

Figure 25. Reassembling the eastern and western triangles times rectangles.

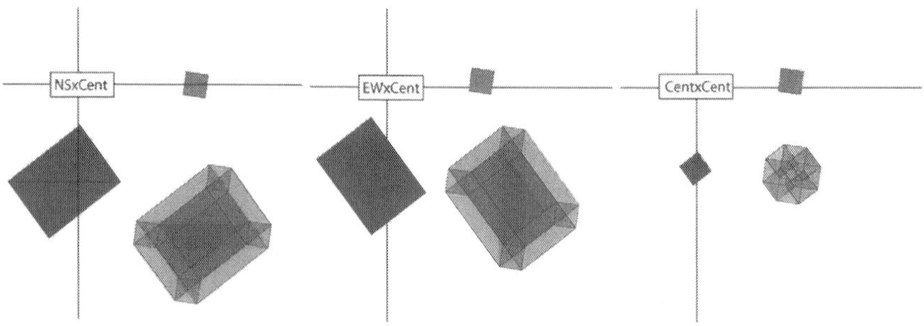

Figure 26. The north/south, east/west and the central square times the center.

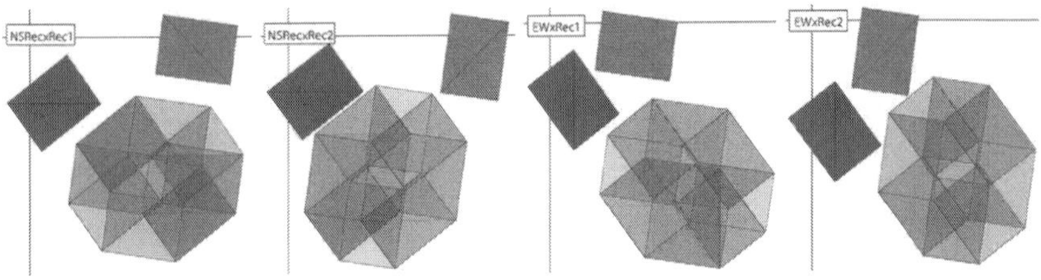

Figure 27. Products of rectangles.

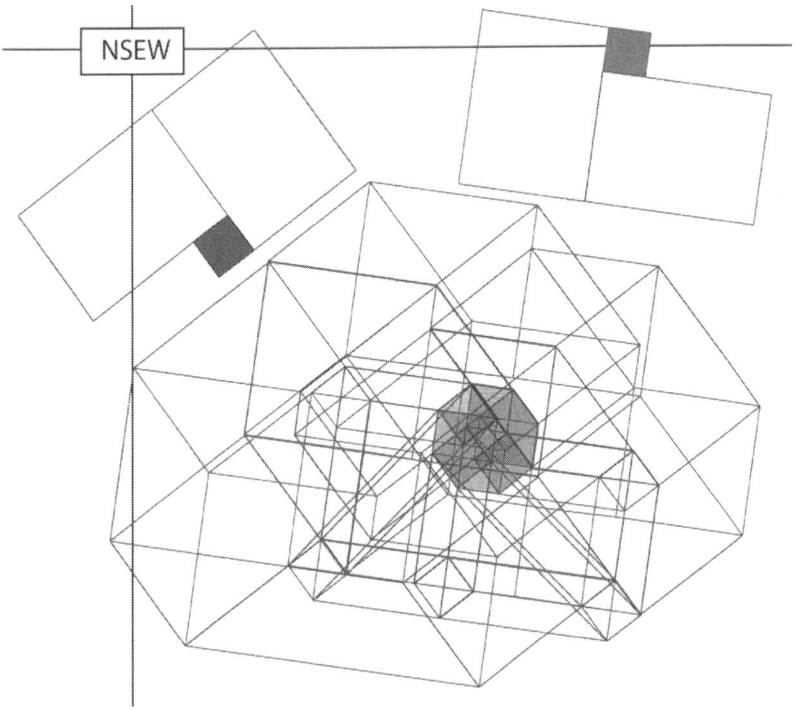

Figure 28. The small central square times itself as a small piece.

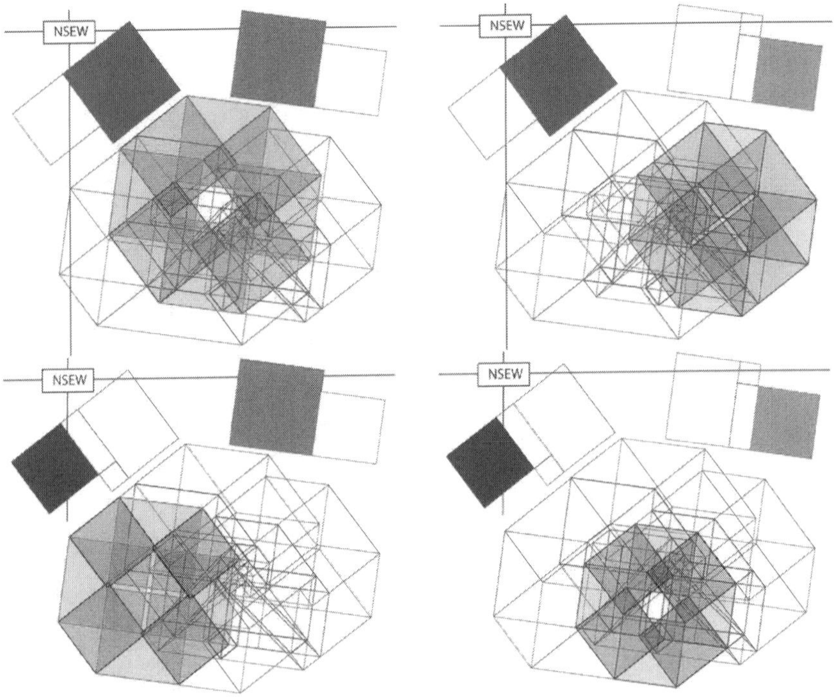

Figure 29. Pieces of the sum of squares by sum of squares.

In this way, we have illustrated that the algebraic identity:

$$z^2w^2 = (x^2 + y^2)(u^2 + v^2) = x^2u^2 + y^2u^2 + x^2v^2 + y^2v^2$$

is realized as a decomposition of the original hyper-rectangle R_{zzww} into the union of four hyper-rectangles R_{xxuu}, R_{yyuu}, R_{xxvv} and R_{yyvv}.

We apply these decompositions to the factors a^4, b^4 and a^2b^2 in the expansion of:

$$(a+b+c)(a+b-c)(a-b+c)(-a+b+c) = 2a^2b^2 + 2a^2c^2 + 2b^2c^2 - a^4 - b^4 - c^4.$$

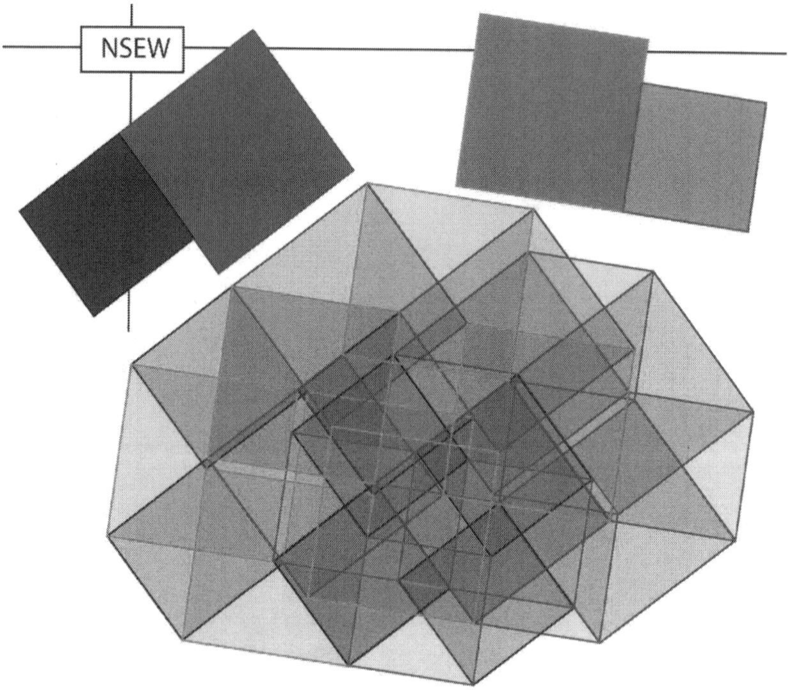

Figure 30. Pieces of the sum of squares by sum of squares.

Here, $a^2 = r^2 + s^2$ and $b^2 = s^2 + (p - r)^2$. In addition, the terms of the form a^2c^2 and b^2c^2 can be reformed by decomposing a^2or b^2 into the five pieces in the proof of the Pythagorean theorem and taking the Cartesian product of each with a square of area $c^2 = p^2$. When the five pieces are reassembled as a union of two squares, the resulting four-dimensional solids are also reassembled.

5. PRODUCTS OF SUMS AND DIFFERENCES OF SQUARES

In Figures 31 and 32, we indicate that products of the form $(x^2-y^2)(x^2-y^2)$ and products of the form $(x^2 + y^2)(z^2 + w^2)$, when thought of as hyper-volumes of four-dimensional solids, can be reassembled into hyper-rectangles whose volumes are the expected terms:

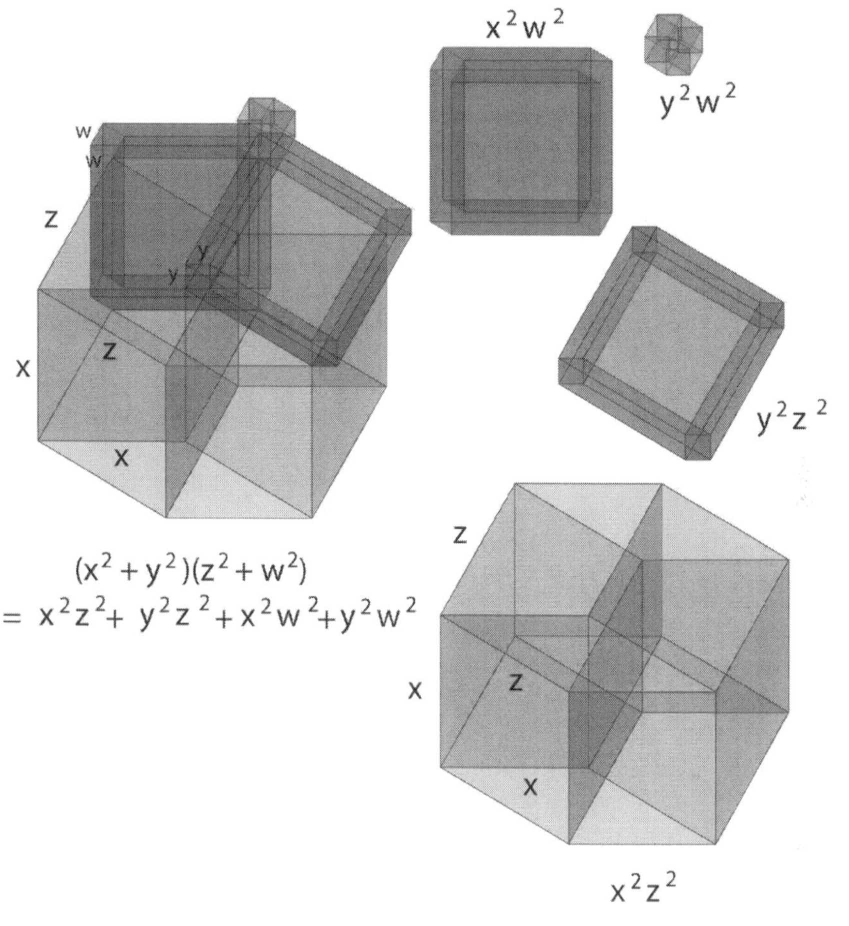

Figure 31. Products of sums of squares.

$$(x^2 - y^2)(x^2 - y^2) = x^4 - 2x^2y^2 + y^4$$

and

$$(x^2 + y^2)(z^2 + w^2) = x^2z^2 + x^2w^2 + y^2z^2 + y^2w^2.$$

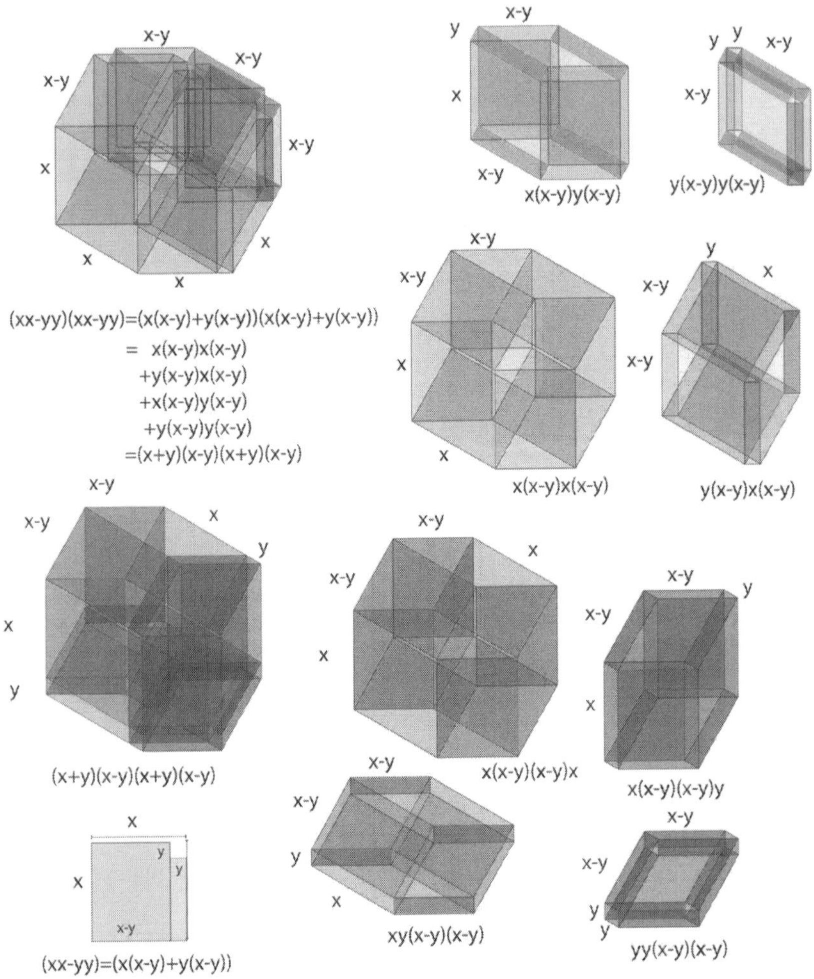

Figure 32. Products of differences of squares.

More specifically, in Figure 31, we have illustrated the two-dimensional projection of a figure from four-space that is constructed by taking the Cartesian product of a planar figure with itself. The planar figure is the union of two squares joined at a vertex. The algebraic expansion above corresponds to the decomposition of this four-dimensional figure into four hyper-rectangles of four-dimensional volumes x^2w^2, y^2w^2, y^2z^2 and x^2z^2.

In the case of the product of a difference of squares with itself, we could use the same figure provided that we interpreted some edge lengths as being negative. However, we can be more descriptive in Figure 32. Start from a planar figure that is constructed by cutting a small square from the northeast corner of a given square, as indicated in the lower left of the figure. Such a figure superficially resembles the geometry of the state of Utah. This figure can be decomposed as the union of two rectangles of size $(x - y) \times x$ and size $y \times (x - y)$. When we take the Cartesian product of such a planar region with itself, it projects to the illustration in the upper left of the corner. Equationally, the expression $(x^2 - y^2)$ is expanded below that illustration. Meanwhile, each term in that expansion is illustrated to the right with the lengths of each edge indicated. In the lower portion of the figure, those same four pieces are illustrated with the regions of sizes $(x - y) \times y \times y \times (x - y)$ and $x \times (x - y) \times (x - y) \times y \times x$ rotated in the set of illustrations to the bottom right. These are then reassembled to form a four-dimensional hyper-solid of size $(x + y) \times (x - y) \times (x + y) \times (x - y)$.

Such multiplications and factorizations were used in the last stages of the original algebraic computation to illustrate that

$$2a^2b^2 + 2a^2c^2 + 2b^2c^2 - a^4 - b^4 - c^4 = 4s^2p^2.$$

6. THE LEFT-HAND SIDE OF THE FORMULA

Consider now the proof that the area of an arbitrary triangle as indicated in Figure 1 is $p \cdot r/2$. The triangle is duplicated and reflected along a horizontal line, and these two copies of the original triangle are joined to create a parallelogram. Thus, twice the area of the triangle is the area of the parallelogram. Meanwhile, the parallelogram is scissors congruent to a rectangle.

If we consider the four-dimensional figure that is a parallelogram-times-parallelogram, then it can be decomposed into four copies of a triangle times a triangle. The decomposition is analogous to that which is depicted in Figure 6, so we do not indicate a new proof. In the former case, the triangles were right triangles. In general, this decomposition can be composed with a linear map from the rectangle-times-rectangle to the parallelogram-times-parallelogram.

In Figure 33, the decomposition of a parallelogram into a right triangle and a trapezoid is lifted into four-space as the decomposition of a parallelogram-times-parallelogram into four pieces: triangle-times-triangle, triangle-times-trapezoid, trapezoid-times-triangle and trapezoid-times-trapezoid. In Figure 34, these same four pieces are reassembled to fill a rectangle-times-rectangle.

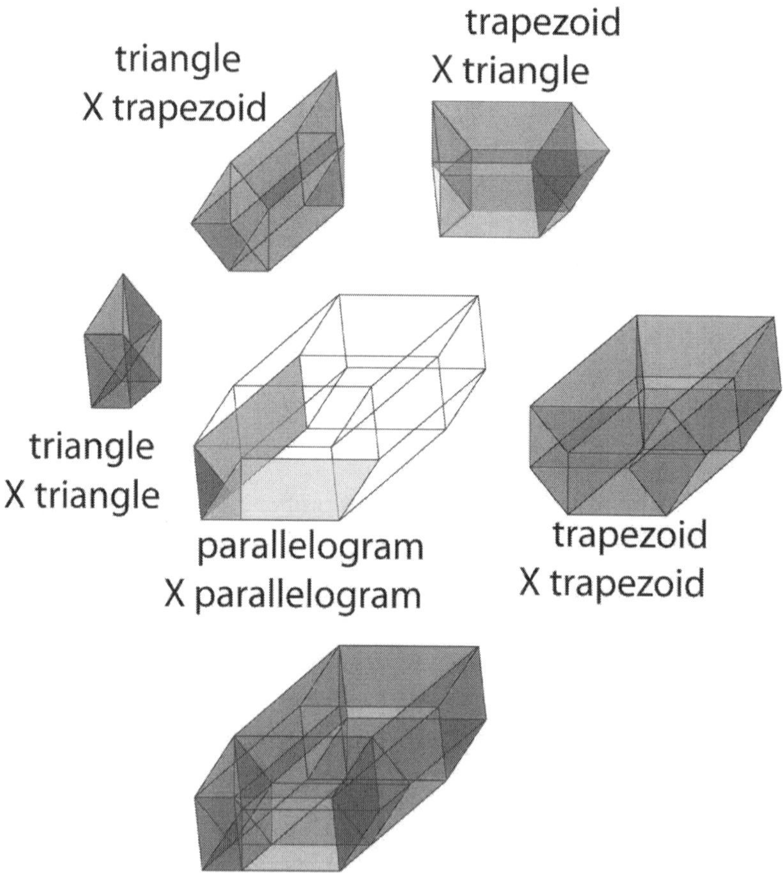

Figure 33. The product of a parallelogram and itself.

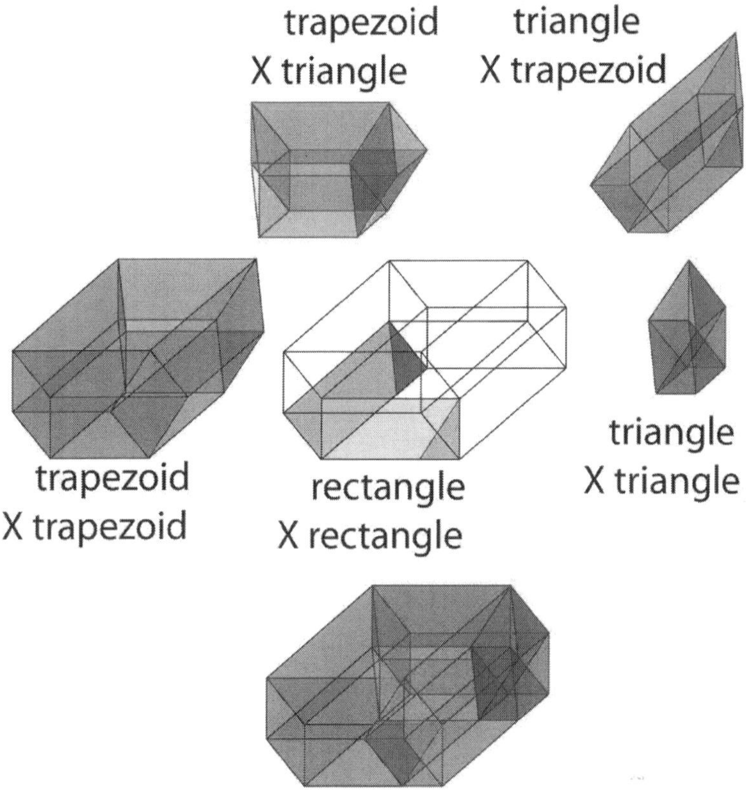

Figure 34. Decomposing a rectangle-times-rectangle into the pieces above.

The left-hand side of Heron's formula is the quantity $4p^2r^2$. Sixteen copies of the original triangle times itself are reassembled into four copies of a rectangle-times-rectangle where the rectangle is of size $p \times r$. Thus, the sequence of equalities:

$$16A^2$$

$$= 4p^2r^2$$

$$= 2a^2b^2 + 2a^2c^2 + 2b^2c^2 - a^4 - b^4 - c^4$$

$$= (a + b + c)(a + b - c)(a - b + c)(-a + b + c).$$

are all realized as scissors congruences on four-dimensional hyper-solids.

Our four-dimensional geometric proof is complete.

7. CONCLUSIONS

Even after more than two and a half centuries of study, many people, including many outstanding mathematicians, have not developed an intuitive grasp of the elements of four-dimensional geometry. Of course, many of us have a more esoteric understanding of higher geometry that is facilitated by our deep understanding of linear algebra and differential forms. Our purpose here has been to demonstrate with two-dimensional projections and with analytic descriptions that many aspects of four-dimensional geometry can be understood. In addition, we have indicated that a number of combinatorial structures and identities are more deeply understood in terms of aspects of higher dimensional geometry.

Therefore, while an algebraic proof of Heron's formula is tedious, yet conceptually simple, a more geometric proof that depends on the geometry of four-spacial dimensions removes some aspects of the tedium. One can envision the pieces fitting tightly together.

Still, there is a cautionary statement to be made. We only perceive two dimensions. Those of us who are sighted see the world projected onto our retinas. As I type these lines, my hands feel the surface of a desk that I presume to be solid. A four-dimensional hyper-solid has a boundary that is three-dimensional. I can only imagine the surfaces of these pieces. As the four-dimensional hyper-solid is projected to three-space, its boundary overlaps in the same way that a ball or orange projects to my eye and appears to be a disk. The front and the back of the ball veil the inner structure.

When we use a temporal metaphor to describe the space-time of the room in which we sit, the authors move within the confines of this $10 \times 11 \times 10$ cubic foot room for a period of, say, one hour. The walls, floor and ceiling of the room form a substantial portion of the boundary as they evolve in time. The"us-then" and the "us-now", that is the interior of the room before the paragraph was written and the interior of the room as the paragraph ends, form the rest of the boundary. Our discussion and our gesticulations in completing this last paragraph and all of the space around us (as a small part of space-time) forms the interior of this hyper-cube of a size of 1100 ft^3/h.

Neither the space-time metaphor nor the projection method produce the full richness of a four-dimensional entity. However, together, they can be used to develop our understanding.

ACKNOWLEDGMENTS

The initial work for this project was conducted when the second author was a UCUR (University Committee for Undergraduate Scholarship) student at the University of South Alabama. It appeared in [8], but we felt that the pictures and our techniques deserved a wider audience. We would like to thank Dror Bar-Natan, Abhijit Champanerkar, Dylan Thurston and the faculty at the University of South Alabama Department of Mathematics and Statistics for valuable conversations.

AUTHOR CONTRIBUTIONS

The authors contributed equally to this work.

CONFLICTS OF INTEREST

The Authors declare no conflict of interest.

REFERENCES

1. Struik, D. *A Concise History of Mathematics*, 4th ed; Dover; Mineola, NY, USA, 1987.

2. *Wikipedia: Heron's Formula*, Available online: http://en.wikipedia.org/wiki/Heron%27s_formula accessed on 28 April 2015.

3. *Wikipedia: Squared Triangular Number*, Available online: http://en.wikipedia.org/wiki/Squared_triangular_n umberaccessed on 28 April 2015.

4. Carter, J.S.; Champanerkar, A. *A geometric method to compute some elementary integrals*, 2006, arXiv, p. math/0608722. [math.HO]. arXiv.org e-Print archive. Available online: http://arxiv.org/pdf/math/ 0608722 accessed on 28 April 2015.

5. Barth, N.R. Computing Cavalieri's Quadrature Formula By a Symmetry of the n-Cube. *Am. Math. Mon.* **2004**, *111*, 811–813.

6. Benko, D. A New approach to Hilbert's Third Problem. *Am. Math. Mon.* **2007**, *114*, 665–676.

7. Dehn, M. Ueber den Rauminhalt. *Math. Annalen* **1901**, *55*, 465–478.

8. Carter, J.S.; Mullens, D.A. *Heron's Formula from a 4-Dimensional Perspective*, Available online: http://www.mi.sanu.ac.rs/vismath/carter2011mart/Carter.pdf accessed on 28 April 2015.

5

CHAPTER

On Links between Rough Sets and Digital Topology

El-Sayed A. Abo-Tabl[1,2]

[1]*Mathematics Department, Faculty of Science, Assiut University, Assiut, Egypt*

[2]*Mathematical Department, Faculty of Science and Arts, Muthnab, Qassim University, Qassim, Saudi Arabia*

ABSTRACT

Rough set theory is a powerful tool for dealing with uncertainty, granularity, and incompleteness of knowledge in information systems. In addition, digital topology deals with properties and features of two-dimensional or three-dimensional digital images that correspond to topological properties of objects. So, we try to describe the relationship between rough sets and digital topology. Firstly, we will study the classifications of topologies in rough sets. Secondly, we will use the upper approximation operator to span the digital line, which is the basic building block of the digital spaces.

Keywords: Rough Sets, Upper Approximation Operator, Topology and Digital Topology

1. INTRODUCTION

The theory of rough sets, proposed by Pawlak [1] [2] , is an extension of the set theory for the study of intelligent systems characterized by insufficient and incomplete information. Using the concepts of the lower and the upper approximations from the rough set theory, knowledge hidden in information systems may be unraveled and expressed in the form of decision rules [2] [3] .

The classical rough approximations are based on equivalence relations, but this requirement is not satisfied in some situations. Thus the classical rough approximations have been extended to the similarity relation based rough sets [4] [5] , the tolerance relation based rough sets [6] , the dominance relation based rough sets [7] , the arbitrary binary relation based rough sets [8] -[10] and the covering-based rough sets [11] -[13] .

An interesting and natural research topic in rough set theory is to study rough set theory via topology. Indeed, Polkowski [14] pointed: topological aspects of rough set theory were recognized early in the framework of topology of partitions. Skowron [15] and Wiweger [16] separately discussed this topic for classical rough set theory in 1988. Polkowski [17] constructed and characterized topological spaces from rough sets based on information systems. Kortelainen [18] considered relationships between modified sets, topological spaces and rough sets based on a pre-order. Skowron et al. [6] [19] generalized the classical approximation spaces to tolerance approximation spaces, and discussed the problems of attribute reduction in these spaces (also see [20]). Lashin et al. [21] introduced the topology generated by a subbase, also defined a topological rough membership function by the subbase of the topology. In addition, connections between fuzzy rough set theory and fuzzy topology were also investigated (see [22] -[24]).

General topology has been considered as the entrance to understanding topology science, moreover the base of general topology is the topological space, which has been considered as a representation of universal space in general, and geometric shape in special, also the mathematical analysis concepts. The general topology has become the appropriated frame for every collection connected by relations. Topology is also a mathematical tool to study information systems and rough sets [15] [21] .

General topology has applications in the theory of image processed by exhibiting algorithms, which apply current knowledge of digital topology (i.e., the study of the geometric and topological properties of digital images [25]). The problems that might arise are, for example finding connected components, set boundaries or any other operations which are needed in image processing. The well-known digital Jordan curve theorem is proved by using topological approach [26] -[28] . The theorem is important in the theory of computer graphics.

The basic building block of the digital n-space is the digital line or the so called Khalimsky line [26] [29] -[31] . This is the set of the integers, Z, equipped with the topology K, generated by $S_K = \{\{2n-1, 2n, 2n+1\} : n \in Z\}$ as a subbase. Thus a set U is open in K if and only if whenever $x \in U$ is an even integer, then $x-1, x+1 \in U$. Nowadays, this topology, called the Khalimsky topology, is one of the most important concepts of the digital topology. It has been studied and used by many authors, e.g., [26] -[37] .

The remainder of this paper is organized as follows. In Section 2, we present the fundamental concepts and properties of Pawlak's rough set theory, general topology and digital topology. Section 3 presents the classifications of topologies in rough sets using the properties of the binary relations. In addition, we use the upper approximation operator to obtain the digital line, which is the basic building block of the digital spaces in Section 4 to try to describe the link between rough sets and digital topology. This paper concludes in Section 5.

2. PRELIMINARIES

In this section, we introduce a review of some basic concepts of topological space and Pawlak approximation space.

A topological space is a pair (U,τ) consisting of a set U and family τ of subset of U satisfying the following conditions:

(T1) $\varnothing \in \tau$ and $U \in \tau$.

(T2) τ is closed under arbitrary union.

(T3) τ is closed under finite intersection.

The pair (U,τ) is called a space, the elements of U are called points of the space, the subsets of U belonging to τ are called open sets in the space, and their complement are called closed sets in the space; the family τ of open subsets of U is also called a topology for U [38] .

Definition 2.1. [39] A topology τ on the set U is called an Alexandroff topology if the intersection of arbitrarily many open sets is still open, or equivalently, the union of arbitrarily many closed sets is still closed.

Definition 2.2. [37] Let U be a nonempty set and cl: $P(U) \to P(U)$ be a function. The space (cl,U) is called Smyth space if the following conditions hold.

1) $cl(\varnothing)=\varnothing$ 2) $cl(A \cup B) \subseteq cl(A) \cup cl(B)$ 3) if $A \subseteq B$ then $cl(A) \subseteq cl(B)$.

Definition 2.3. For the set of integer Z we have:

(1) In Z the 2-neighbors of x is $N_2(x)=\{x-1,x+1\}$ (2) In Z^2 the 4-neighbors of (x,y) is $N_4((x,y))=\{(x-1,y),(x+1,y),(x,y-1),(x,y+1)\}$,

(3) In Z^2 the 8-neighbors of (x,y) is

$$N_8((x,y))=\{(x-1,y-1),(x-1,y),(x-1,y+1),(x+1,y-1),(x+1,y),(x+1,y+1),(x,y-1),(x,y+1)\}.$$

Definition 2.4. Let X be a nonempty set. Then the following are defined.

1) A relation R on a set X is reflexive if and only if xRx for all x in X.

2) A relation R on a set X is symmetric if and only if xRy implies yRx.

3) A relation R on a set X is transitive if xRy, yRz, then xRz.

A relation R on a set X is called an equivalence relation if it is reflexive, symmetric and transitive.

Definition 2.5. [2] Let R be an equivalence relation on a nonempty set U. For any subset $A \subseteq U$, the lower and upper approximations of A according to R are then defined as

$$\underline{R}(A)=\{x \in U:[x]_R \subseteq A\}$$

$\overline{R}(A)=\{x \in U:[x]_R \cap A \neq \varnothing\}$ where $[x]_R$ is called an equivalence class of $x \in U$.

Yao [37] extended the Pawlak's definitions of a rough set to any binary relation R as follows:

Definition 2.6. For the pair (U,R), the set xR is defined as $xR=\{y \in U:xRy\}$ called the right neighborhood of an element $x \in U$.

Definition 2.7. Let R be any binary relation on a nonempty set U. For any subset $A \subseteq U$, the lower and the upper approximations of A according to R are then defined as $\underline{R}(A)=\{x \in U:xR \subseteq A\}$ $\overline{R}(A)=\{x \in U:xR \cap A \neq \varnothing\}$

Obviously, if R is an equivalence relation, $xR=[x]_R$ and these definitions are equivalent to the original Pawlak's definitions.

We list the properties that are of interest in the theory of rough sets, let $A,B \subseteq U$:

L_1. $\overline{R}(A)=(\overline{R}(A^C))^C$, where A^C denotes the complementation of A in U.

L2. $\underline{R}(U) = U$.

L3. $\underline{R}(A \cap B) = \underline{R}(A) \cap \underline{R}(B)$.

L4. $\underline{R}(A \cup B) \supseteq \underline{R}(A) \cup \underline{R}(B)$.

L5. $A \subseteq B \Rightarrow \underline{R}(A) \subseteq \underline{R}(B)$.

L6. $\underline{R}(\varnothing) = \varnothing$.

L7. $\underline{R}(A) \subseteq A$.

L8. $A \subseteq \underline{R}(\overline{R}(A))$.

L9. $\underline{R}(A) \subseteq \underline{R}(\underline{R}(A))$.

L10. $\overline{R}(A) \subseteq \underline{R}(\overline{R}(A))$.

U1. $\overline{R}(A) = (\overline{R}(A^C))^C$.

U2. $\overline{R}(\varnothing) = \varnothing$.

U3. $\overline{R}(A \cup B) = \overline{R}(A) \cup \overline{R}(B)$.

U4. $\overline{R}(A \cap B) \subseteq \overline{R}(A) \cap \overline{R}(B)$.

U5. $A \subseteq B \Rightarrow \overline{R}(A) \subseteq \overline{R}(B)$.

U6. $\overline{R}(U) = U$.

U7. $A \subseteq \overline{R}(A)$.

U_8. $\overline{R}(\underline{R}(A)) \subseteq A$.

U_9. $\overline{R}(\overline{R}(A)) \subseteq \overline{R}(A)$.

U_{10}. $\overline{R}(\underline{R}(A)) \subseteq \underline{R}(A)$.

Properties L_1 and U_1 state that two approximations are dual to each other. Hence, properties with the same numbers may be regarded as dual properties. Properties L_9, L_{10}, U_9 and U_{10} are expressed in terms of set inclusion. The standard version using set equality can be derived from L_1- L_{10} and U_1 - U_{10}. For example, it follows from L_7 and L_9 that $\underline{R}(A) = \underline{R}(\underline{R}(A))$. It should also be noted that these properties are not independent.

We can find the basic axioms for extended topological spaces in Table 1.

3. TOPOLOGY CLASSIFICATIONS BASED ON ROUGH SETS

In this section, we discuss the classification of topologies in the theory of rough sets.

Proposition 3.1. For any relation R on a nonempty set U and for every $A, B \subseteq U$ the properties L_1 - L_5 and U_1 - U_5 hold according to Definition 2.7.

Proof. See [9] .

Example 3.1. Let $R = \{(a,a),(b,a),(b,c),(b,d),(c,d),(d,c)\}$ be a relation on a nonempty set $U = \{a,b,c,d\}$. Then $aR = \{a\}$, $bR = \{a,c,d\}$, $cR = \{d\}$ and $dR = \{c\}$.

Then $\tau = \{A : R(A) = A\} = \{\emptyset, U, \{a\}, \{c,d\}\}$ is not a topology on U.

Theorem 3.1. Suppose that $(U,\underline{R},\overline{R})$ is a rough approximation space. If $(U,\underline{R},\overline{R})$ satisfies the properties L_1 - L_5 and U_1 - U_5, then $(U,\overline{R})$ is a Smyth space [37] .

Proposition 3.2. For any reflexive relation R on a nonempty set U and for every $A,B \subseteq U$ the properties L_1 - L_7 and U_1 - U_7 hold according to Definition 2.7.

Proof. See [9] .

Example 3.2. Let $R=\{(a,a),(b,b),(c,c),(d,d),(a,b),(c,d),(d,a)\}$ be a reflexive relation on a nonempty set $U=\{a,b,c,d\}$.
Then $aR=\{a,b\},bR=\{b\},cR=\{c,d\}$ and $dR=\{a,d\}$.
Then $\tau=\{\varnothing,U,\{b\},\{a,b\},\{a,b,d\}\}$ is a topology on U.

Theorem 3.2. If $(U,\underline{R},\overline{R})$ satisfies the properties L_1 - L_7 and U_1 - U_7, then $(U,\overline{R})$ is a pre-topology or Čech closure space.

Proposition 3.3. For any reflexive and symmetric relation R on a nonempty set U and for every $A,B \subseteq U$ the properties L_1 - L_8 and U_1 - U_8 hold according to Definition 2.7.

Proof. See [9] .

Example 3.3. Let $R=\{(a,a),(b,b),(c,c),(d,d),(a,b),(b,a),(b,c),(c,b)\}$ be a reflexive and symmetric relation on a nonempty set $U=\{a,b,c,d\}$.
Then $aR=\{a,b\},bR=\{a,b,c\},cR=\{b,c\}$ and $dR=\{d\}$.
Then $\tau=\{\varnothing,U,\{d\},\{a,b,c\}\}$ is a topology on U in which each open (closed) set is also closed (open) set.

Theorem 3.3. If $(U,\underline{R},\overline{R})$ satisfies the properties L_1 - L_8 and U_1 - U_8, then $(U,\overline{R})$ is a pre-topology or Čech closure space in which each open (closed) set is also closed (open).

Proposition 3.4. For any reflexive and transitive relation R on a nonempty set U and for every $A,B \subseteq U$ the properties L_1 - L_7, L_9, U_1 - U_8 and U_9 hold according to Definition 2.7.

Proof. See [9] .

Example 3.4. Let $R=\{(a,a),(b,b),(c,c),(d,d),(a,b),(b,d),(a,d),(a,c),(c,d)\}$ be a reflexive and transitive relation on a nonempty set $U=\{a,b,c,d\}$. Then $aR=\{a,b,c,d\},bR=\{b,d\},cR=\{c,d\}$ and $dR=\{d\}$. Then $\tau=\{\varnothing,U,\{d\},\{b,d\},\{c,d\},\{b,c,d\}\}$ is an Alexandroff topology on U.

Birkhoff [40] showed that there exists a one-to-one correspondence between the set of all reflexive and transitive binary relations and Alexandroff topologies.

Theorem 3.4. If $(U,\underline{R},\overline{R})$ satisfies the properties L_1 - L_7, L_9, U_1 - U_7 and U_9, then $(U,\overline{R})$ is an Alexandroff topological space.

Proposition 3.5. For any equivalence relation R on a nonempty set U and for every $A,B \subseteq U$ the properties L_1 - L_{10} and U_1 - U_{10} hold according to Definition 2.7.

Proof. See [9] .

Table 1. The basic axioms for extended topological spaces.

Axioms	Property	Name
K_0	L_2 U_2	Normality Co-normality
K_1	L_3 U_3	Multiplication Addition
K_2	L_5 U_5	Monotone Monotone
K_3	L_7 U_7	Counteraction Extension
K_4	L_8 U_8	Symmetry Symmetry
K_5	L_9 U_9	Idempotent Idempotent
K_6	L_{10} U_{10}	Super-Symmetry Super-Symmetry

Example 3.5. Let $R=\{(a,a),(b,b),(c,c),(d,d),(a,b),(b,a),(b,c),(c,b),(a,c),(c,a)\}$ be an equivalence relation on a nonempty set $U=\{a,b,c,d\}$. Then $aR=bR=cR=\{a,b,c\}$ and $dR=\{d\}$.
Then $\tau=\{\varnothing,U,\{d\},\{a,b,c\}\}$ is a topology on U in which each open (closed) set is also closed (open) set, and socalled clo-open topology.

Theorem 3.5. If $(U,\underline{R},\overline{R})$ satisfies the properties L_1 - L_{10} and U_1 - U_{10}, then $(U,\overline{R})$ is an Alexandroff topological space in which each open (closed) set is also closed (open) set, and so-called clo-open space.

We summarize these results in Table 2, which indicates the generalized topological spaces depending on the properties of rough sets.

4. CONNECTING DIGITAL TOPOLOGY AND ROUGH SETS

In this section, we try to get the structure of digital topology from the concepts of the theory of rough sets (the upper approximate operator).

Definition 4.1. Let R be any relation on a set of integer Z. We define another relation R' on Z as follows:

$$xR'y \Leftrightarrow x \in \overline{R}(y)$$

on the other hand, we can write the relation R′ in an equivalent form as follows:

$$xR'y \Leftrightarrow x \rightarrow y$$

Theorem 4.1. 1) The relation R on Z is reflexive if and only if R′ is also reflexive.

2) The relation R on Z is transitive if and only if R′ is also transitive.

Proof. 1) Let R be a reflexive relation on $Z \Leftrightarrow x \in R(x) \Leftrightarrow xR'x \Leftrightarrow R'$ is reflexive.

2) Assume that xR′y and yR′z, then $x \in \overline{R}(y)$ and $y \in \overline{R}(z) \Rightarrow xRy$ and yRz, hence by transitivity of R we get xRz, thus $x \in \overline{R}(z)$, that implies xR′z. Therefore R′ is transitive. Conversely, suppose that xRy and yRz, then $x \in \overline{R}(y)$ and $y \in \overline{R}(z)$, that implies xR′y and yR′z, hence by transitivity of R′ we get xR′z, thus $x \in \overline{R}(z)$, so that xRz. Therefore R is transitive.

The following conditions hold on Z using the relation →1) if two points $x, y \in Z$ are 2-neighbors, then either x → y or y → x2) if two points $x, y \in Z$ are not 2-neighbors, then both x → y and y → x not hold.

Lemma 4.1. If $x_1, x_2, x_3 \in Z$ such that $x_2 \in N_2(x_1)$ and $x_3 \in N_2(x_2)$, then we have either $x_1 \to x_2 \leftarrow x_3$ or $x_1 \leftarrow x_2 \to x_3$.

Proof. By condition (1) we have either $x_1 \to x_2$ or $x_2 \to x_1$. Let $x_1 \to x_2$. By condition (1) either $x_2 \to x_3$ or $x_3 \to x_2$. If $x_2 \to x_3$, then by transitivity $x_1 \to x_3$ which contradiction to condition (2). Hence $x_3 \to x_2$.

Then by Lemma 4.1 we can draw the graph of the relation → as follows (Figure 1).

Note that, in the line → ○ ← is equivalent to the open set containing the element "○" and ← □ → is equivalent to the closed set containing the element "□", also the digital line is an Alexandroff topology.

Now, if we consider R is a relation on Z^2, then the following conditions hold:

a) If two points $x, y \in Z^2$ are 4-neighbors, then either x → y or y → x,

Table 2. Generalized topological spaces defining axioms are indicated by*.

Generalized topologies/Axioms	K_0	K_1	K_2	K_3	K_4	K_5	K_6
Smyth space	*	*	*				
Pre-topology	*	*	*	*			
Pre-topology with every open set is closed	*	*	*	*	*		
Alexandroff topology	*	*	*	*		*	
Clo-open topology	*	*	*	*	*	*	*

b) If two points $x, y \in Z^2$ are not 8-neighbors, then both x → y and y → x not hold.

Lemma 4.2. If x_1, x_2, x_3 which are in the same horizontal or vertical gride line of Z2 such that $x_2 \in N_4(x_1)$ and $x_3 \in N_4(x_2)$, then we have either $x_1 \to x_2 \leftarrow x_3$ or $x_1 \leftarrow x_2 \to x_3$ on a gride line.

Proof. By condition a) we have either $x_1 \to x_2$ or $x_2 \to x_1$. Let $x_1 \to x_2$. Also, by condition a) either $x_2 \to x_3$ or $x_3 \to x_2$. If $x_2 \to x_3$, then by transitivity $x_1 \to x_3$ which contradiction to condition b). Hence $x_3 \to x_2$.

In the plane note that, the element "○" is equivalent to the open-open, the element "□" is equivalent to the closed-closed, and the element "●" is called the mixed point and equivalent to the open-closed or the closedopen.

We must note that, there are two topologies on Z^2 which satisfy the conditions 1 and 2, one of them was described in Khalimsky et al. [26] and the other was introduced by Marcus et al. [41] .

Figure 2 describe the topology in [26] . Furthermore, if there is no mixed point, then we have the topology in [41] , which described in Figure 3.

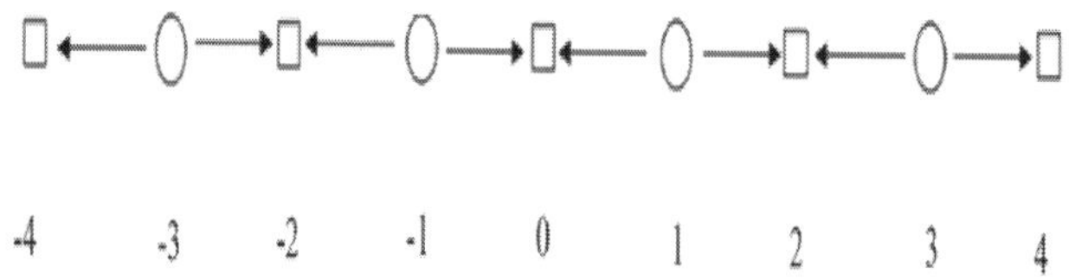

Figure 1. Khalimsky line.

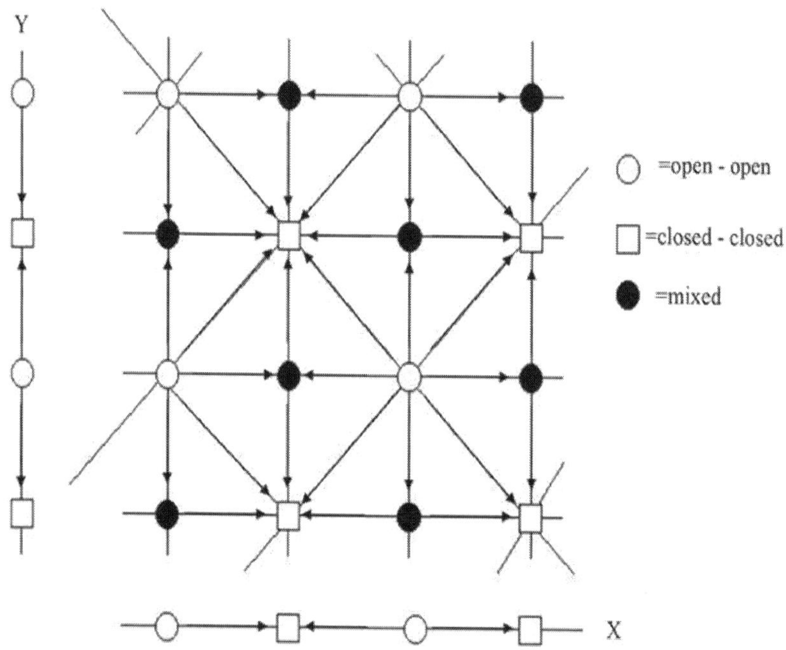

Figure 2. Khalimsky topology in the plane.

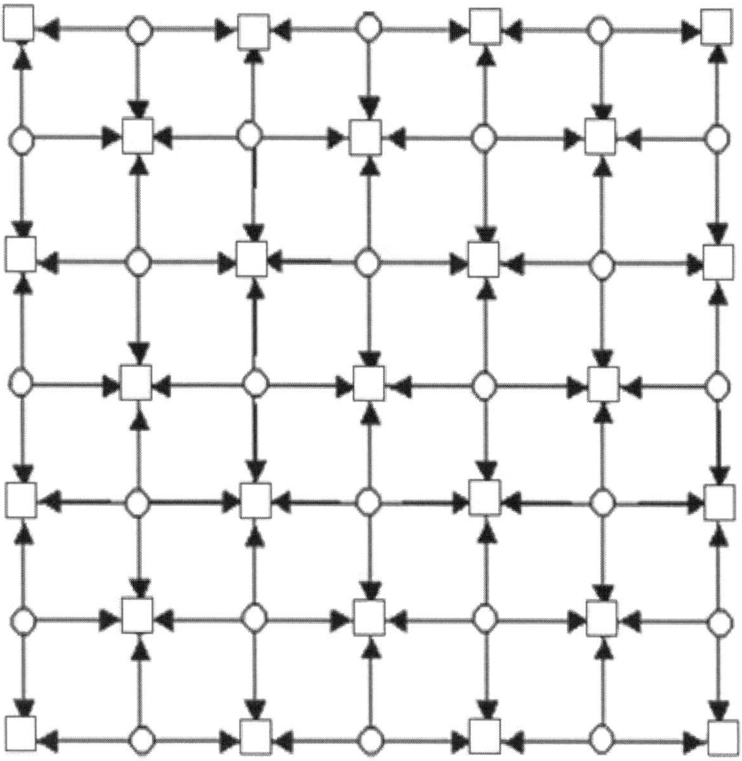

Figure 3. Marcus topology in the plane.

5. CONCLUSIONS

In short, topology is a branch of mathematics, whose concepts are fundamental not only to all branches of mathematics, but also in real life applications. Image plays important role in real life. In the past the process of image analysis took place via various mathematical models with an acceptable amount of error. Digital topology is a new accurate approach.

So we use the theory of rough sets to generate the digital spaces, in order to narrow the gap between rough sets and digital topology.

REFERENCES

1. Pawlak, Z. (1982) Rough Sets. International Journal of Computer and Information Science, 11, 341-356.

2. Pawlak, Z. (1991) Rough Sets: Theoretical Aspects of Reasoning about Data. Kluwer Academic Publishers, Boston.

3. Kryszkiewicz, M. (1998) Rough Set Approach to Incomplete Information Systems. Information Sciences, 112, 39-49.

4. Abo-Tabl, E.A. (2013) Rough Sets and Topological Spaces Based on Similarity. International Journal of Machine Learning and Cybernetics, 4, 451-458.

5. Slowinski, R. and Vanderpooten, D. (2000) A Generalized Definition of Rough Approximations Based on Similarity. IEEE Transactions on Knowledge and Data Engineering, 12, 331-336.

6. Skowron, A. and Stepaniuk, J. (1996) Tolerance Approximation Spaces. Fundamenta Informaticae, 27, 245-253.

7. Zhang, H., Ouyang, Y. and Wangc, Z. (2009) Note on Generalized rough Sets Based on Reflexive and Transitive Relations. Information Sciences, 179, 471-473.

8. Liu, G.L. and Zhu, W. (2008) The Algebraic Structures of Generalized Rough Set Theory. Information Sciences, 178, 4105-4113.

9. Yao Y.Y. (1998) Relational Interpretations of Neighborhood Operators and Rough Set Approximation Operators. Information Sciences, 111, 239-259.

10. Zhu, W. (2009) Relationship between Generalized Rough Sets Based on Binary Relation and Covering. Information Sciences, 179, 210-225.

11. Liu, G.L. and Sai, Y. (2009) A Comparison of Two Types of Rough Sets Induced by Coverings. International Journal of Approximate Reasoning, 50, 521-528.

12. Zhu, W. (2007) Topological Approaches to Covering Rough Sets. Information Sciences, 177, 1499-1508.

13. Zhu, W. (2009) Relationship among Basic Concepts in Covering-Based Rough Sets. Information Sciences, 179, 2478-2486.

14. Polkowski, L. (2002) Rough Sets: Mathematical Foundations. Physica-Verlag, Heidelberg.

15. Skowron, A. (1988) On Topology in Information System. Bulletin of Polish Academic Science and Mathematics, 36, 477-480.

16. Wiweger, A. (1988) On Topological Rough Sets. Bulletin of the Polish Academy of Sciences Mathematics, 37, 51-62.

17. Polkowski, L. (2001) On Fractals Defined in Information Systems via Rough Set Theory. Proceedings of the RSTGC-2001. Bulletin International Rough Set Society, 5, 163-166.

18. Kortelainen J. (1994) On Relationship between Modified Sets, Topological Spaces and Rough Sets. Fuzzy Sets and Systems, 61, 91-95.

19. Skowron, A., Swiniarski, R. and Synak, P. (2005) Approximation Spaces and Information Granulation. Transactions on Rough Sets III: Lecture Notes in Computer Science, 3400, 175-189.

20. Jarvinen, J. and Kortelainen, J. (2007) A Unifying Study between Model-Like Operators, Topologies, and Fuzzy Sets. Fuzzy Sets and Systems, 158, 1217-1225.

21. Lashin, E., Kozae, A., Khadra, A.A. and Medhat, T. (2005) Rough Set Theory for Topological Spaces. International Journal of Approximate Reasoning, 40, 35-43.

22. Li, T.J., Yeung, Y. and Zhang, W.X. (2008) Generalized Fuzzy Rough Approximation Operators Based on Fuzzy Covering. International Journal of Approximate Reasoning, 48, 836-856.

23. Qin, K. and Pei, Z. (2005) On the Topological Properties of Fuzzy Rough Sets. Fuzzy Sets and Systems, 151, 601-613.

24. Srivastava, A.K. and Tiwari, S.P. (2003) On Relationships among Fuzzy Approximation Operators, Fuzzy Topology, and Fuzzy Automata. Fuzzy Sets and Systems, 138, 197-204.

25. Rosenfeld, A. (1979) Digital topology. The American Mathematical Monthly, 86, 621-630.

26. Khalimsky, E.D., Kopperman, R. and Meyer, P.R. (1990) Computer Graphics and Connected Topologies on Finite Ordered Sets. Topology and Its Applications, 36, 1-17.

27. Kiselman, C.O. (2000) Digital Jordan Curve Theorems. Discrete Geometry for Computer Imagery (DGCI), 1953, 46-56.

28. Slapal, J. (2013) Topological Structuring of the Digital Plane. Discrete Mathematics and Theoretical Computer Science, 15, 165-176.

29. Khalimsky, E.D., Kopperman, R. and Meyer, P.R. (1990) Boundaries in Digital Planes. Journal of Applied Mathematics and Stochastic Analysis, 3, 27-55.

30. Kong, T.Y., Kopperman, R. and Meyer, P.R. (1991) A Topological Approach to Digital Topology. The American Mathematical Monthly, 98, 901-917.

31. Kopperman, R., Meyer, P.R. and Wilson, R.G. (1991) A Jordan Surface Theorem for Three-Dimensional Digital Spaces. Discrete and Computational Geometry, 6, 155-161.

32. Bai, Y., Han, X. and Prince, J.L. (2009) Digital Topology on Adaptive Octree Grids. Journal of Mathematical Imaging and Vision, 34, 165-184.

33. Eckhardt, U. and Latecki, L.J. (2003) Topologies for the Digital Spaces Z^2 and Z^3. Computer Vision and Image Understanding, 90, 295-312.

34. Melin, E. (2007) Digital Surfaces and Boundaries in Khalimsky Spaces. Journal of Mathematical Imaging and Vision, 28, 169-177.

35. Slapal, J. (2003) Closure Operations for Digital Topology. Theoretical Computer Science, 305, 457-471.

36. Slapal, J. (2006) Digital Jordan Curves. Topology and Its Application, 153, 3255-3264.

37. Smyth, M.B. (1995) Semi-Metrics, Closure Spaces and Digital Topology. Theoretical Computer Science, 151, 257-276.

38. Sierpenski, W. and Krieger, C. (1956) General Topology. University of Toronto, Toronto.

39. Alexandroff, P. (1937) Diskrete Raume. Matematicheskii Sbornik, 2, 501-519.

40. Birkhoff, G. (1937) Rings of Sets. Duke Mathematical Journal, 3, 383-548.

41. Marcus, D., Wyse, F., et al. (1970) A Special Topology for the Integers (Problem 5712). The American Mathematical Monthly, 77, 1119.

6

CHAPTER

The Quantum sl_2-Invariant of a Family of Knots

Abdul Rauf Nizami, Mobeen Munir, Malka Shah Bano

Division of Science and Technology, University of Education, Lahore, Pakistan

ABSTRACT

We give a general formula of the quantum sl_2-invariant of a family of braid knots. To compute the quantum invariant of the links we use the Lie algebra $g = sl_2$ in its standard two-dimensional representation. We also recover the Jones polynomial of these knots as a special case of this quantum invariant.

Keywords: Quantum Invariant; Jones Polynomial; Braid Knot

1. INTRODUCTION

The discovery of the Jones polynomial inspired many people to search for other skein relations compatible with Reidemeister moves and thus defined knot polynomials. This led to the introduction of the HOMFLY and Kauffmans polynomials. It soon became clear that all these polynomials are the first members of a vast family of knot invariants called quantum invariants.

The original idea of quantum invariants was proposed by E. Witten in [1]. Witten's approach coming from physics was not completely justified from the mathematical viewpoint. The first mathematically definition of quantum invariants of links and 3-manifolds was given by Reshetikhin and Turaev [2,3], who used in their construction the notion of quantum groups introduced shortly before that by V. Drinfeld in [4] (see also [5]) and M. Jimbo in [6]. In fact, a quantum group is not a group at all. Instead, it is a family of algebras, more precisely, of Hopf algebras, depending on a complex parameter q and satisfying certain axioms. The quantum group $U_q\mathfrak{g}$ of a semisimple Lie algebra $\mathfrak{g}$ is a remarkable deformation of the universal enveloping algebra of $\mathfrak{g}$ (corresponding to the value $q = 1$) in the class of Hopf algebras.

This paper is organized as follows: In Section 2, we give the basic ideas about knots, tangles, the Jones polynomial, Lie algebra representations, and construction of quantum invariants. In Section 3, we present the main result along with its specialization to the Jones polynomial.

2. PRELIMINARY NOTIONS

2.1. Basic Concepts of Knots

A knot is a circle embedded in $\mathbb{R}^3$. Knots are usually studied via projecting them on a plan; a projection with extra information of overcrossing and undercrossing is called the knot diagram

A crossing trivial knot trefoil knot

A crossing trivial knot trefoil knot Two knots are called isotopic if one of them can be transformed to the other by a diffeomorphism of the ambient space $\square^3$ onto itself. A fundamental result about the isotopic knot diagrams is:

Two unoriented knots K_1 and K_2 are equivalent if and only if a diagram of K_1 can be transformed into a diagram of K_2 by a finite sequence of ambient isotopies of the plane and three local (Reidemeister) moves:

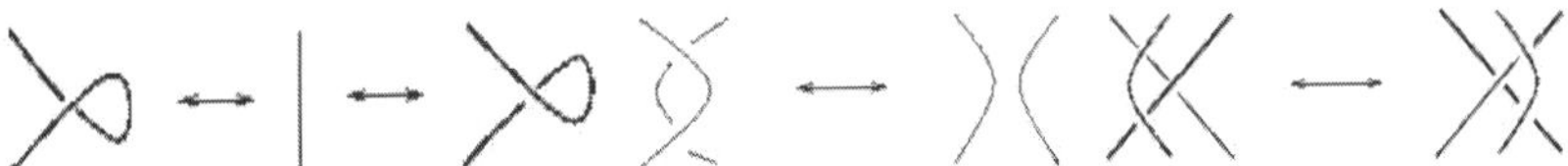

R1 R2 R3 The set of all knots that are equivalent to a knot K is called a class of K. By a knot K we shall always mean a class of the knot K.

The main question of knot theory is Which two links are equivalent and which are not? To address this question one needs a knot invariant, a function that gives one value on all knots that belong to a single class and gives different values (but not always) on knots that belong to different classes. The present work is concerned with this question.

2.2. Tangles

A tangle is a generalization of a knot which at the same time is simpler and more complicated than a knot. On one hand, knots are a particular case of tangles, on the other hand, knots can be represented as combinations of (simple) tangles.

A tangle in a knot projection is a region in the projection plane surrounded by a circle such that the knot crosses the circle exactly at four places.

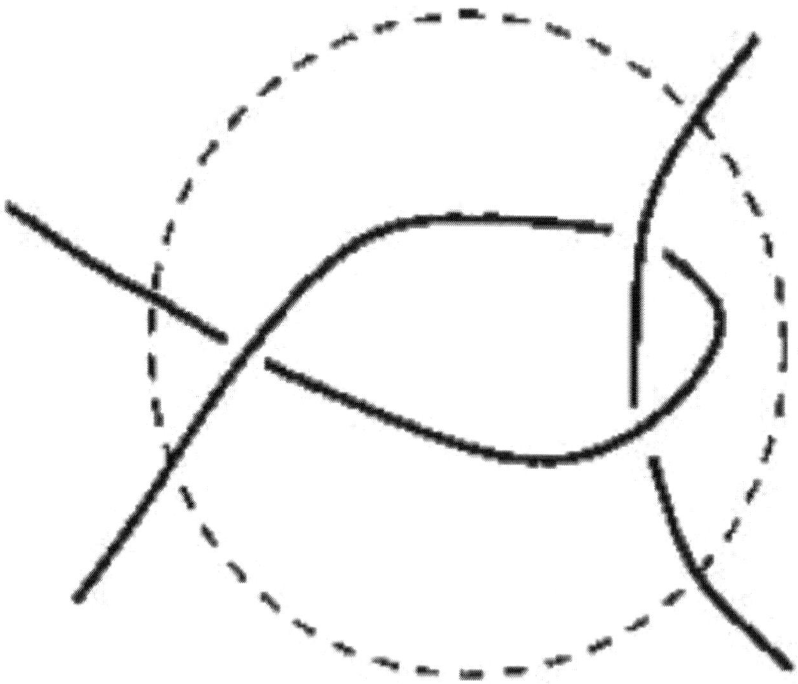

A tangle

A tangle The following two operations are defined on tangles: When the bottom of a tangle T_1 coincides with the top of another tangle T_2, the product $T_1 \cdot T_2$ is defined by putting T_1 on top of T_2. (For oriented tangles we also require the consistency of orientations.)

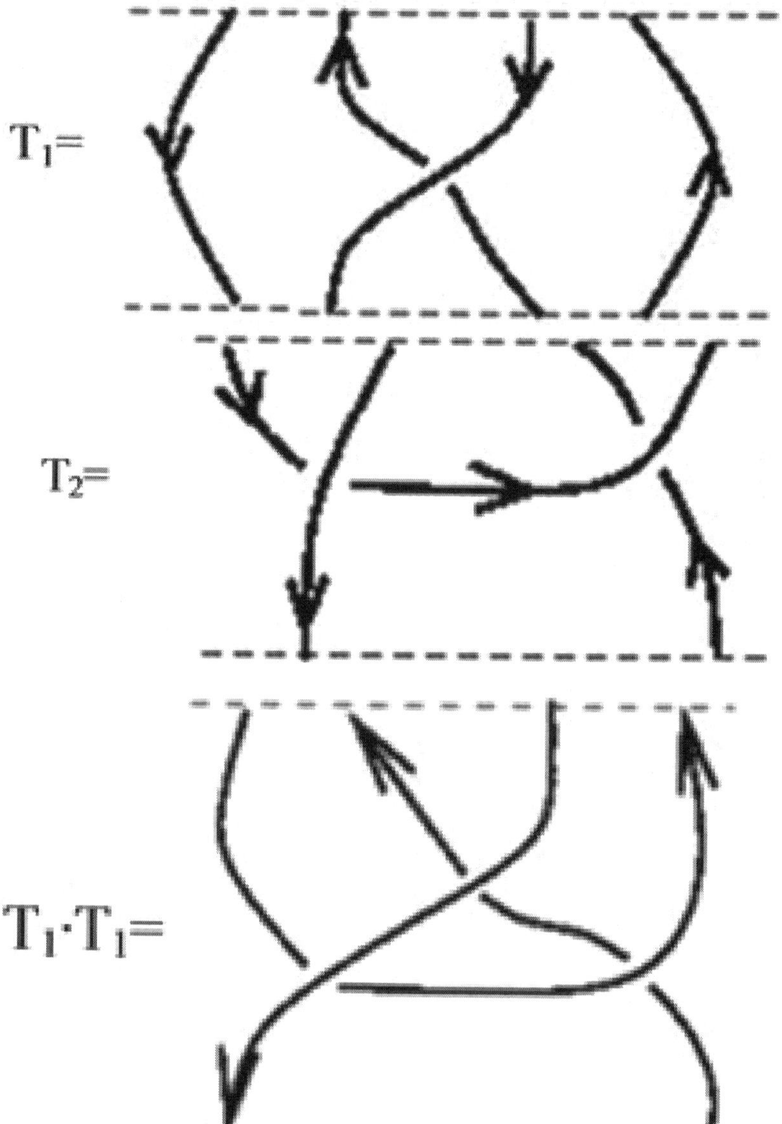

The second operation, tensor product, is defined by placing one tangle next to the other tangle (of the same height).

$T_1 \otimes T_2 =$ 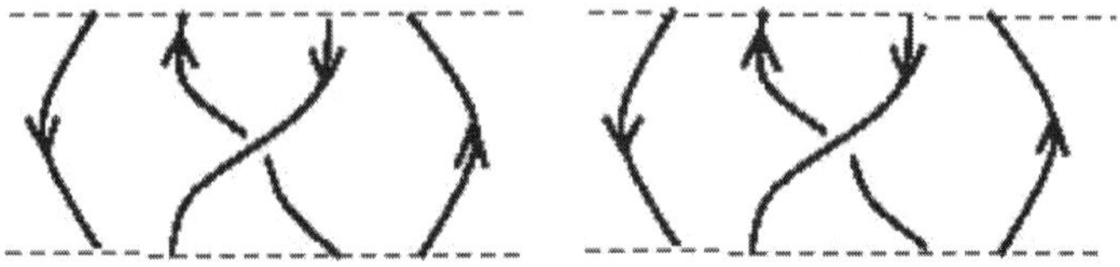

2.3. The Jones Polynomial

In 1985, V. F. R. Jones revolutionized knot theory by defining the Jones polynomial as a knot invariant via Von Neumann algebras [7]. However, in 1987 L. H. Kauffman introduced in [8] a state-sum model construction of the Jones polynomial that was purely combinatorial and remarkably simple; we follow this approach.

Definition 1 [7-9] The Jones polynomial $V_K(t)$ of an oriented link K is a Laurent polynomial in the variable $\sqrt{t}$ satisfying the skein relation

$$t^{-1}V_{K_+}(t) - tV_{K_-}(t) = \left(t^{1/2} - t^{-1/2}\right)V_{K_0}(t),$$

and that the value of the unknot is 1. Here K_+, K_-, and K_0 are three oriented links having diagrams that are isotopic everywhere except at one crossing where they differ as in the figure below:

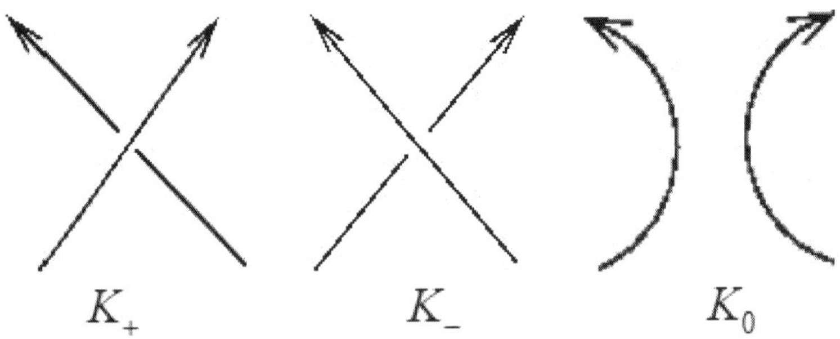

K_+ K_- K_0

For instance, it is easy to verify that the Jones polynomial of the left-handed trefoil knot (which is denoted by 3_1 in the knot table) is V = $-t^{-4}+t^{-3}+t^{-1}$.

2.4. Lie Algebra Representations

Let $\mathfrak{g}$ be a semisimple Lie algebra and let V be its finite-dimensional representation. One can view V as a representation of the universal enveloping algebra $U(\mathfrak{g})$. It is remarkable that this representation can also be deformed with parameter q to a representation of the quantum group $U_q\mathfrak{g}$. The vector space V remains the same, but the action now depends on q. For a generic value of q all irreducible representations of $U_q\mathfrak{g}$ can be obtained in this way. However, when q is a root of unity the representation theory is different and resembles the representation theory of q in finite characteristic. It can be used to derive quantum invariants of 3-manifolds. For the purposes of knot theory it is enough to use generic values of q, that is, those which are not roots of unity.

An important property of quantum groups is that every representation gives rise to a solution R of the quantum Yang-Baxter equation

$$(R\otimes id_V)\circ(id_V\otimes R)\circ(R\otimes id_V)=(id_V\otimes R)\circ(R\otimes id_V)\circ(id_V\otimes R)$$

where R (the R-matrix) is an invertible linear operator $R:V\otimes V\rightarrow V\otimes V$, and both sides of the equation are understood as linear transformations $V\otimes V\otimes V\rightarrow V\otimes V\otimes V$.

In case of Lie algebra $\mathfrak{g}=\mathfrak{sl}_2$ and its standard two-dimensional representation, the R-matrix has the form

$$R = \begin{cases} e_1 \otimes e_1 \mapsto q^{1/4} e_1 \otimes e_1 \\ e_1 \otimes e_2 \mapsto q^{-1/4} e_2 \otimes e_1 \\ e_2 \otimes e_1 \mapsto q^{-1/4} e_1 \otimes e_2 + \left(q^{1/4} - q^{-3/4}\right) e_2 \otimes e_1 \\ e_2 \otimes e_2 \mapsto q^{1/4} e_2 \otimes e_2 \end{cases}$$

for an appropriate basis $\{e_1, e_2\}$ of the space V. The inverse of R is

$$R^{-1} = \begin{cases} e_1 \otimes e_1 \mapsto q^{-1/4} e_1 \otimes e_1 \\ e_1 \otimes e_2 \mapsto q^{1/4} e_2 \otimes e_1 + \left(-q^{3/4} + q^{-1/4}\right) e_1 \otimes e_2 \\ e_2 \otimes e_1 \mapsto q^{1/4} e_1 \otimes e_2 \\ e_2 \otimes e_2 \mapsto q^{-1/4} e_2 \otimes e_2 \end{cases}$$

2.5. Construction of Quantum Invariants

The general procedure of constructing quantum invariants is as follows (see details in [10]). Consider a knot diagram in the plane and take a generic horizontal line. To each intersection point of the line with the diagram assign either the representation space V or its dual V^* depending on whether the orientation of the knot at this intersection is directed upwards or downwards. Then take the tensor product of all such spaces over the whole horizontal line. If the knot diagram does not intersect the line, then the corresponding vector space is the ground field □ .

A portion of a knot diagram between two such horizontal lines represents a tangle T. We assume that this tangle is framed by the black board framing. With T we associate a linear transformation $\theta^{fr}(T)$ from the vector space corresponding to the bottom of T to the vector space corresponding to the top of T. The following three properties hold for the linear transformation $\theta^{fr}(T)$:

- $\theta^{fr}(T)$ is an invariant of the isotopy class of the framed tangle T;
- $\theta^{fr}(T_1 \cdot T_2) = \theta^{fr}(T_1) \circ \theta^{fr}(T_2)$;
- $\theta^{fr}(T_1 \otimes T_2) = \theta^{fr}(T_1) \otimes \theta^{fr}(T_2)$.

$$
\begin{array}{c}
V^* \otimes V \otimes V \otimes V \\
\theta^{fr}(T_1) \uparrow \\
V \otimes V^* \otimes V \otimes V \\
\theta^{fr}(T_2) \uparrow \\
V \otimes V
\end{array}
$$

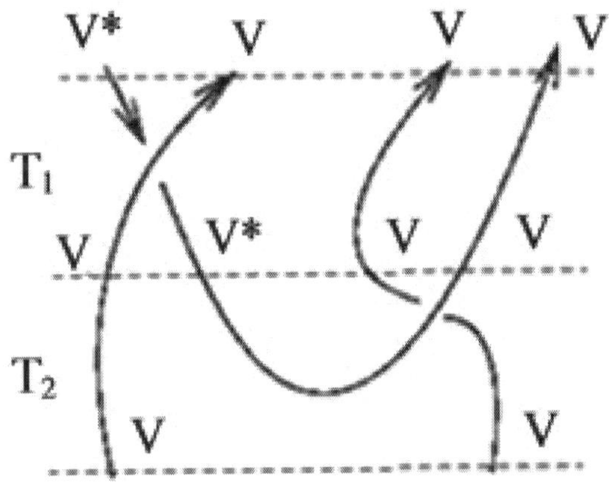

Now we can define a knot invariant $\theta^{fr}(K)$ regarding the knot K as a tangle between the two lines below and above K. In this case $\theta^{fr}(K)$ would be a linear transformation from C to C. Since our linear transformations depend on the parameter q, this number is actually a function of q.

Because of the multiplicity property $\theta^{fr}(T_1 \cdot T_2) = \theta^{fr}(T_1) \circ \theta^{fr}(T_2)$ it is enough to define $\theta^{fr}(T)$ only for elementary tangles T such as a crossing, a minimum or a maximum point. This is precisely where quantum groups come in. Given a quantum group $U_q g$ and its finite-dimensional representation V, one can associate certain linear transformations with elementary tangles in a way consistent with the Turaev oriented moves [11]. The R-matrix appears here as the linear transformation corresponding to a positive crossing. Of course, for a trivial tangle consisting of a single string connecting the top and bottom, the corresponding linear operator should be the identity transformation. So we have the following correspondence valid for all quantum groups:

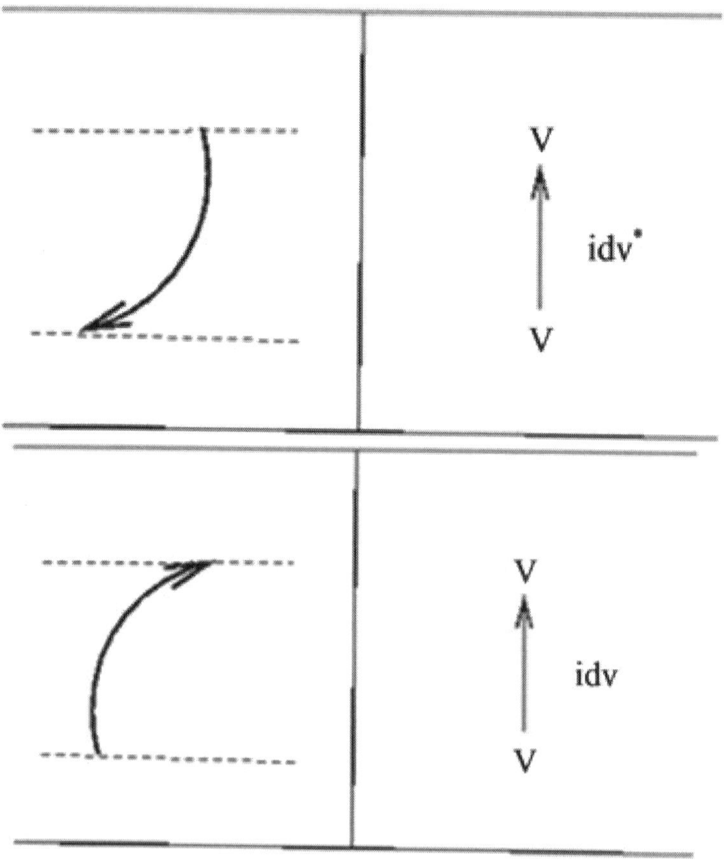

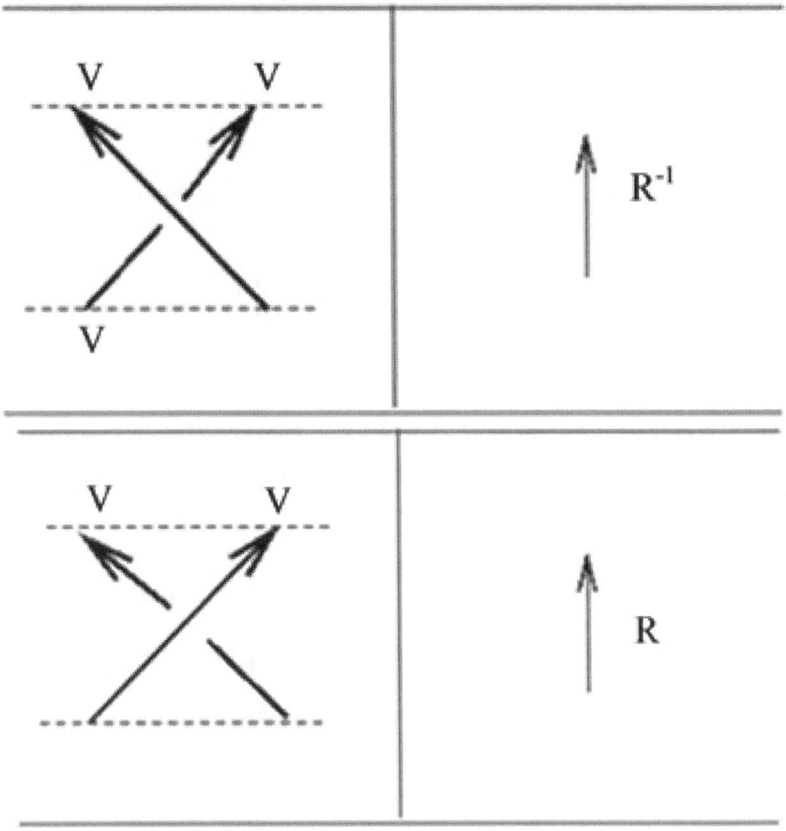

Using this one can verify that θ^{fr} remains invariant under all three Reidemeister moves, for details see [11].

To complete the construction of our quantum invariant we should assign appropriate operators to the minimum and maximum points. These depend on all the data involved: the quantum group, the representation and the R-matrix. For the quantum group $U_q\mathrm{sl}_2$, its standard two-dimensional representation V and the R- matrix, these operators are:

$$\begin{array}{cc} V^{*}\otimes V & q^{-1/2}e^{1}\otimes e_{1}+q^{1/2}e^{2}\otimes e_{2} \\ \uparrow & \uparrow \\ \square & 1 \end{array}$$

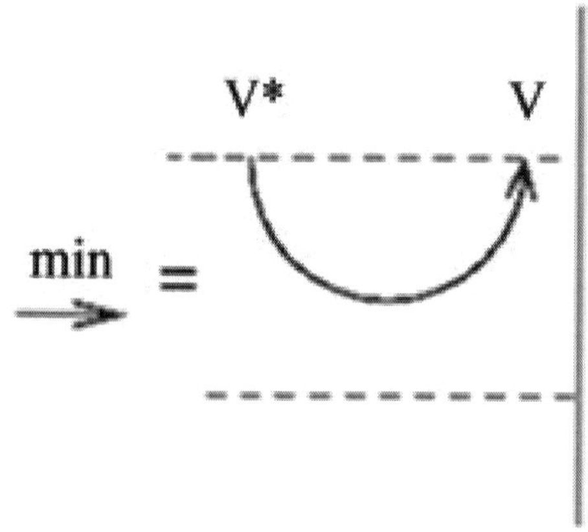

$$\begin{array}{cc} V\otimes V^{*} & e_{1}\otimes e^{1}+e_{2}\otimes e^{2} \\ \uparrow & \uparrow \\ \square & 1 \end{array}$$

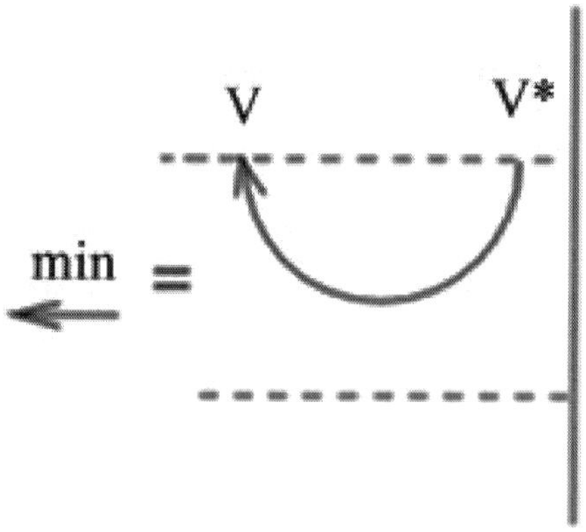

$$\begin{array}{ccccc} \square & q^{1/2} & 0 & 0 & q^{-1/2} \\ \uparrow & \uparrow & \uparrow & \uparrow & \uparrow \\ V \otimes V^{*} & e_1 \otimes e^1 & e_1 \otimes e^2 & e_2 \otimes e^1 & e_2 \otimes e^2 \end{array}$$

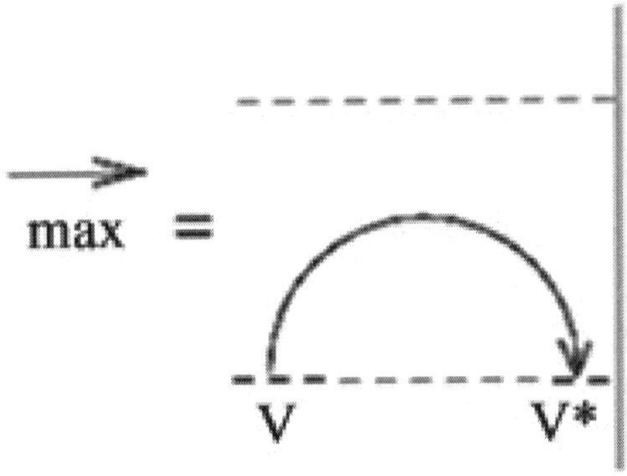

$$\begin{array}{ccccc} \square & 1 & 0 & 0 & 1 \\ \uparrow & \uparrow & \uparrow & \uparrow & \uparrow \\ V^{*} \otimes V & e^1 \otimes e_1 & e^1 \otimes e_2 & e^2 \otimes e_1 & e^2 \otimes e_2 \end{array}$$

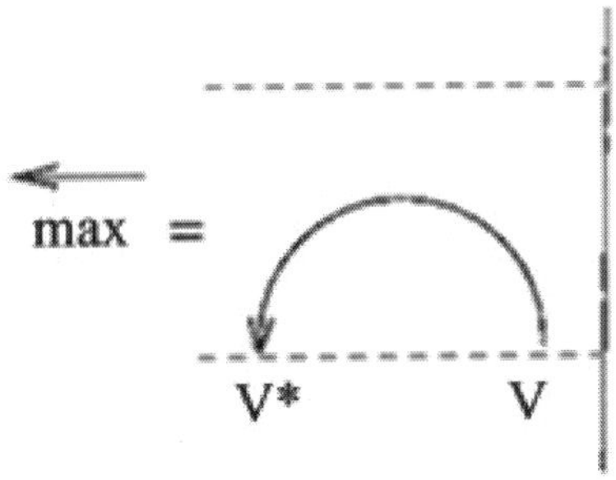

where $\{e^1, e^2\}$ is the basis of V^{*}, dual to the basis $\{e_1, e_2\}$ of V.

In the following example we compute the quantum sl_2-invariant for the unknot.

Example 1 Let us compute the sl_2-quantum invariant of the unknot. Represent the unknot as a product of two tangles and compute the composition of the corresponding transformations.

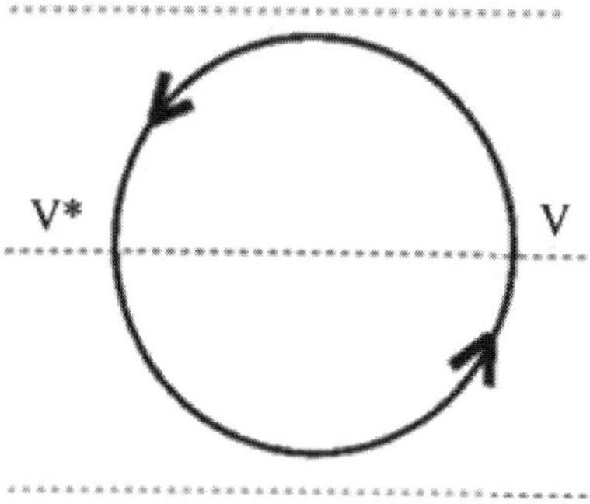

$$\begin{array}{cc} \square & \underbrace{q^{-1/2}} + \underbrace{q^{1/2}} \\ \uparrow & \uparrow \\ V^* \otimes V & \underbrace{\overbrace{q^{-1/2} e^1 \otimes e_1} + \overbrace{q^{1/2} e^2 \otimes e_2}} \\ \uparrow & \uparrow \\ \square & 1 \end{array}$$

So, $\theta^{fr}(\text{unknot}) = q^{1/2} + q^{-1/2}$. Therefore, in order to normalize our invariant so that its value on the unknot is equal to 1, we must divide $\theta^{fr}(\cdot)$ by $q^{1/2} + q^{-1/2}$, and denote this normalized invariant by $\tilde{\theta}^{fr}(\cdot)$. (We shall write the precise formula for $\tilde{\theta}^{fr}(\cdot)$ in the main result.)

3. MAIN RESULT

Here we give the general formulas of the quantum sl_2-invariants of the braid knot $\widehat{x_1^n}$ for odd n.

Proposition 1 The quantum invariant of $\widehat{x_1^n}$, when n is odd, is

$$\tilde{\theta}\left(\widehat{x_1^n}\right) = -q^{\frac{3n-1}{2}} + q^{\frac{3n-3}{2}} - q^{\frac{3n-5}{2}} + \cdots + q^{\frac{n+3}{2}} + q^{\frac{n-1}{2}}. \tag{1.1}$$

Proof 1 We prove it by induction on n.

For $n = 1$, we receive the following braid knot along with its tensor product.

$$\begin{array}{c}
\square \\
\uparrow \\
V^{*} \otimes V \\
\uparrow \\
V^{*} \otimes V \otimes V \otimes V^{*} \\
\uparrow id_{V^{*}} \otimes R^{-1} \otimes id_{V^{*}} \\
V^{*} \otimes V \otimes V \otimes V^{*} \\
\uparrow \\
V \otimes V^{*} \\
\uparrow \\
\square
\end{array}$$

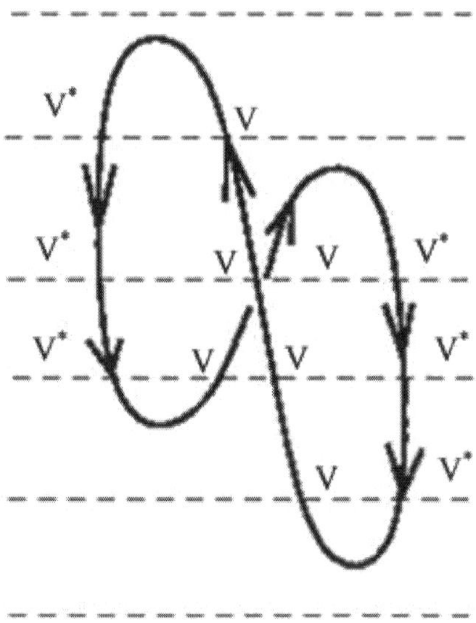

Note that the map $\square \to V \otimes V^*$ sends $1 \in \square$ into the tensor $e_1 \otimes e^1 + e_2 \otimes e^2$. Also, the map $V \otimes V^* \to V^* \otimes V \otimes V \otimes V^*$ sends $e_1 \otimes e^1 + e_2 \otimes e^2$ into the tensor

$$q^{-1/2} e^1 \otimes e_1 \otimes e_1 \otimes e^1 + q^{-1/2} e^1 \otimes e_1 \otimes e_2 \otimes e^2 + q^{1/2} e^2 \otimes e_2 \otimes e_1 \otimes e^1 + q^{1/2} e^2 \otimes e_2 \otimes e_2 \otimes e^2$$

.

Now applying R^{-1} to each middle factor, we get

$$q^{-1/2} e^1 \otimes \left(q^{-1/4} e_1 \otimes e_1\right) \otimes e^1 + q^{-1/2} e^1 \otimes \left(q^{1/4} e_2 \otimes e_1 + \left(-q^{3/4} + q^{-1/4}\right) e_1 \otimes e_2\right) \otimes e^2$$

$$+ q^{1/2} e^2 \otimes \left(q^{1/4} e_1 \otimes e_2\right) \otimes e^1 + q^{1/2} e^2 \otimes \left(q^{-1/4} e_2 \otimes e_2\right) \otimes e^2$$

$$= q^{-1/2} e^1 \otimes e_1 \otimes q^{-1/4} e_1 \otimes e^1 + q^{-1/2} e^1 \otimes e_2 \otimes q^{1/4} e_1 \otimes e^2$$

$$+ q^{-1/2} e^1 \otimes e_1 \otimes \left(-q^{3/4} + q^{-1/4}\right) e_2 \otimes e^2 + q^{1/2} e^2 \otimes e_1 \otimes q^{1/4} e_2 \otimes e^1 + q^{1/2} e^2 \otimes e_2 \otimes q^{-1/4} e_2 \otimes e^2$$

Finally, the two maps at the top contract the whole tensor into the linear transformation

$$\theta^{fr}\left(\square_{x_1}\right) = q^{-1/2} q^{-1/4} q^{1/2} + q^{-1/2}\left(-q^{3/4} + q^{-1/4}\right) q^{-1/2} + q^{1/2} q^{-1/4} q^{-1/2} = q^{-1/4} + q^{-5/4}.$$

Hence the unframed normalized $\mathfrak{sl}_2$ -quantum invariant of $\square_{x_1}$ is

$$\tilde{\theta}\left(\square_{x_1}\right) = q^{\frac{-3wr(K)}{4}} \left[\frac{\theta^{fr}\left(\square_{x_1}\right)}{q^{1/2} + q^{-1/2}}\right] = q^{\frac{-3/2(-1)}{2}} \left[q^{1/4} + q^{-3/4} - q^{1/4}\right]$$

$$= q^{3/4}\left[q^{1/4} + q^{-3/4} - q^{1/4}\right] = q + q^0 - q = 1.$$

To get a clear picture, we also compute the quantum invariant of the knots x_1^3 (which is actually the left trefoil) and x_1^5. First of all, we proceed for x_1^3 :

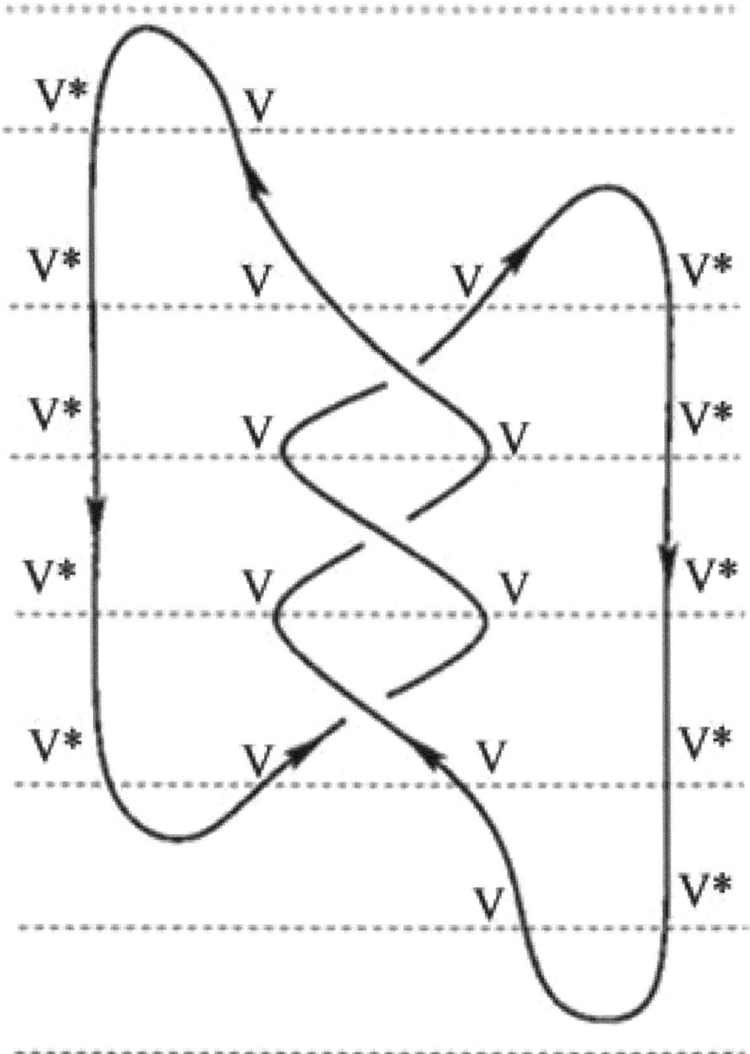

$$
\begin{array}{l}
\Box \\
\uparrow \\
V^{*} \otimes V \\
\uparrow \\
V^{*} \otimes V \otimes V \otimes V^{*} \\
\uparrow id_{V^{*}} \otimes R^{-1} \otimes id_{V^{*}} \\
V^{*} \otimes V \otimes V \otimes V^{*} \\
\uparrow id_{V^{*}} \otimes R^{-1} \otimes id_{V^{*}} \\
V^{*} \otimes V \otimes V \otimes V^{*} \\
\uparrow \\
V \otimes V^{*} \\
\uparrow \\
\Box
\end{array}
$$

The map at the bottom sends $1 \in \Box$ into the tensor $e_1 \otimes e^1 + e_2 \otimes e^2$.

Now the map $V \otimes V^{*} \to V^{*} \otimes V \otimes V \otimes V^{*}$ sends the above tensor into the tensor

$$q^{-1/2} e^1 \otimes e_1 \otimes e_1 \otimes e^1 + q^{-1/2} e^1 \otimes e_1 \otimes e_2 \otimes e^2 + q^{1/2} e^2 \otimes e_2 \otimes e_1 \otimes e^1 + q^{1/2} e^2 \otimes e_2 \otimes e_2 \otimes e^2.$$

Then applying R^{-3} to two tensor factors in the middle we get

$$q^{-1/2}e^1 \otimes \left(q^{-3/4}e_1 \otimes e_1\right) \otimes e^1$$
$$+q^{-1/2}e^1 \otimes \left[\left(-q^{9/4}+q^{5/4}-q^{1/4}+q^{-3/4}\right)e_1 \otimes e_2 + \left(-q^{7/4}-q^{3/4}-q^{-1/4}\right)e_2 \otimes e_1\right] \otimes e^2$$
$$+q^{1/2}e^2 \otimes \left[\left(q^{7/4}-q^{3/4}+q^{-1/4}\right)e_1 \otimes e_2 + \left(-q^{5/4}+q^{1/4}\right)e_2 \otimes e_1\right] \otimes e^1$$
$$+q^{1/2}e^2 \otimes \left(q^{-3/4}e_2 \otimes e_2\right) \otimes e^2.$$

Finally, the two maps at the top contract the whole tensor into a number

$$\begin{aligned}\theta^{fr}(3_1) &= q^{-1/2}q^{-3/4}q^{-1/2} + q^{-1/2}\left(-q^{9/4}+q^{5/4}-q^{1/4}+q^{-3/4}\right)q^{-1/2} \\ &\quad + q^{1/2}\left(-q^{5/4}+q^{1/4}\right)q^{1/2} + q^{1/2}q^{-3/4}q^{-1/2} \\ &= 2q^{-3/4} - q^{5/4} + q^{1/4} - q^{-3/4} + q^{-7/4} - q^{9/4} + q^{5/4} \\ &= q^{-7/4} + q^{-3/4} + q^{1/4} - q^{9/4}\end{aligned}$$

Dividing by the normalizing factor $q^{1/2}+q^{-1/2}$ we get

$$\frac{\theta^{fr}(3_1)}{q^{1/2}+q^{-1/2}} = q^{-5/4} + q^{3/4} - q^{7/4}.$$

The invariant $\theta^{fr}(K)$ remains unchanged under the second and third reidemeister moves. However it varies under the first reidemeister move and thus depends on the framing. One can deframe it, that is, manufacture an invariant of unframed knots out of it, according to the formula

$$\theta(K)=q^{\frac{-c\cdot w(K)}{2}}\theta^{fr}(K),$$

where $w(K)$ is the writhe of the knot diagram and c is the quadratic Casimir number defined by the Lie algebra $\mathfrak{g}$ and its representation. For $\mathfrak{sl}_2$ and the standard 2-dimensional representation $c=\frac{3}{2}$. Since the writhe of the left trefoil is -3, the unframed normalized quantum invariant is

$$\tilde{\theta}(3_1)=\frac{\theta(3_1)}{q^{1/2}+q^{-1/2}}=q^{9/4}\left(q^{-5/4}+q^{3/4}-q^{7/4}\right)=q+q^3-q^4.$$

This can be further written as

$$\tilde{\theta}\left(\overset{\square^{3}}{X_1}\right)=-q^{\frac{3(3)-1}{2}}+q^{\frac{3(3)-3}{2}}+q\left[\tilde{\theta}\left(\overset{\square^{1}}{X_1}\right)\right]. \tag{1.2}$$

For $n=5$, the knot and the corresponding tensor products are:

$$
\begin{array}{l}
\square \\
\uparrow \\
V^* \otimes V \\
\uparrow \\
V^* \otimes V \otimes V \otimes V^* \\
\uparrow id_{V^*} \otimes R^{-1} \otimes id_{V^*} \\
V^* \otimes V \otimes V \otimes V^* \\
\uparrow id_{V^*} \otimes R^{-1} \otimes id_{V^*} \\
V^* \otimes V \otimes V \otimes V^* \\
\uparrow id_{V^*} \otimes R^{-1} \otimes id_{V^*} \\
V^* \otimes V \otimes V \otimes V^* \\
\uparrow id_{V^*} \otimes R^{-1} \otimes id_{V^*} \\
V^* \otimes V \otimes V \otimes V^* \\
\uparrow id_{V^*} \otimes R^{-1} \otimes id_{V^*} \\
V^* \otimes V \otimes V \otimes V^* \\
\uparrow \\
V \otimes V^* \\
\uparrow \\
\square
\end{array}
$$

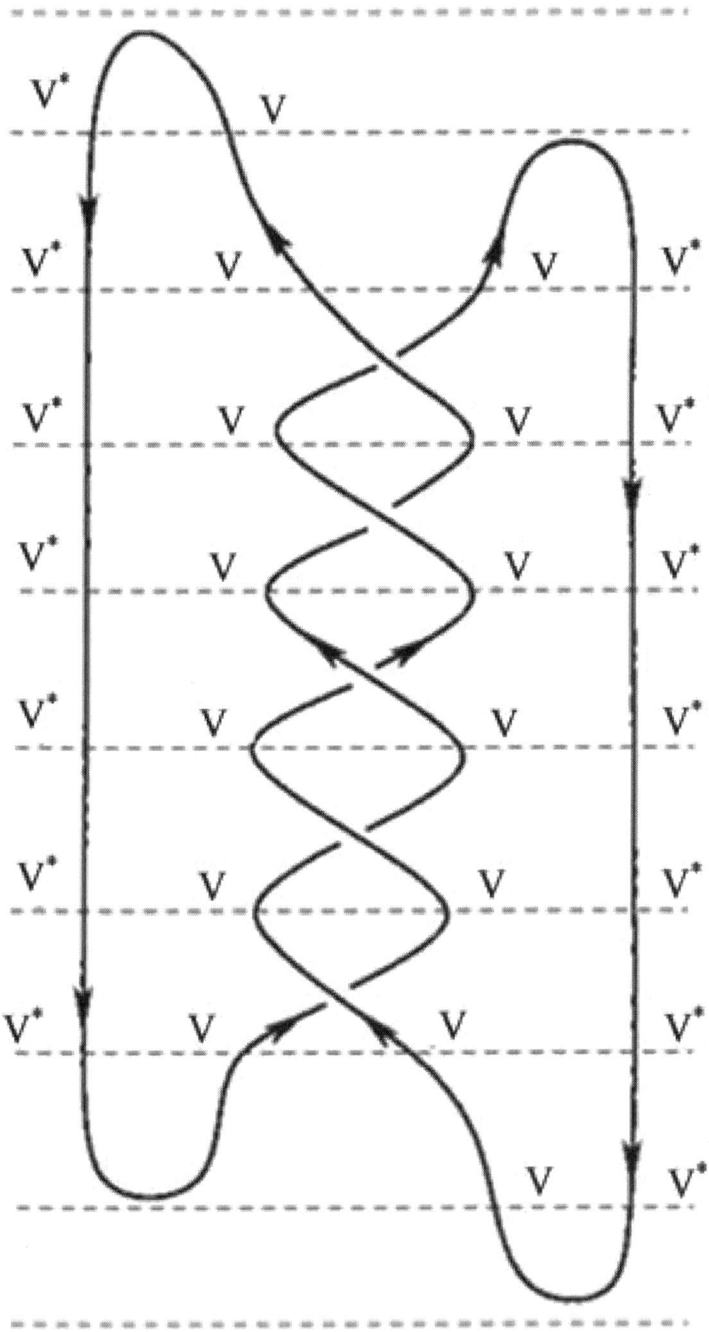
V*
V
V*
V
V
V*
V*
V
V
V*
V*
V
V
V*
V*
V
V
V*
V*
V
V
V*
V*
V
V
V*
V
V*

With some computations, similar to the computations of $\tilde{\theta}\left(\square^{3}_{x_1}\right)$, we get

$$\tilde{\theta}\left(\square^{5}_{x_1}\right) = -q^7 + q^6 - q^5 + q^4 + q^2 = -q^{\frac{3(5)-1}{2}} + q^{\frac{3(5)-3}{2}} + q\left[\tilde{\theta}\left(\square^{3}_{x_1}\right)\right]. \tag{1.3}$$

Similarly,

$$\tilde{\theta}\left(\square^{7}_{x_1}\right) = -q^{\frac{3(7)-1}{2}} + q^{\frac{3(7)-3}{2}} + q\left[\tilde{\theta}\left(\square^{5}_{x_1}\right)\right]. \tag{1.4}$$

We now assume the result (1.1) holds for $n = k$, that is

$$\tilde{\theta}\left(\square^{k}_{x_1}\right) = -q^{\frac{3k-1}{2}} + q^{\frac{3k-3}{2}} - q^{\frac{3k-5}{2}} + \cdots + q^{\frac{k+3}{2}} + q^{\frac{k-1}{2}} \tag{1.5}$$

Now for $n = k+2$ we have

$$\begin{aligned}
\tilde{\theta}\left(\square^{k+2}_{x_1}\right) &= -q^{\frac{3(k+2)-1}{2}} + q^{\frac{3(k+2)-3}{2}} + q\left(\tilde{\theta}\left(\square^{k}_{x_1}\right)\right) \\
&= -q^{\frac{3(k+2)-1}{2}} + q^{\frac{3(k+2)-3}{2}} + q\left[-q^{\frac{3k-1}{2}} + q^{\frac{3k-3}{2}} - q^{\frac{3k-5}{2}} + q^{\frac{3k-7}{2}} + \cdots + q^{\frac{k+3}{2}} + q^{\frac{k-1}{2}}\right] \\
&= -q^{\frac{3(k+2)-1}{2}} + q^{\frac{3(k+2)-3}{2}} - q^{\frac{3(k+2)-5}{2}} + q^{\frac{3(k+2)-7}{2}} + \cdots + q^{\frac{(k+2)+3}{2}} + q^{\frac{(k+2)-1}{2}},
\end{aligned}$$

and the proof is finished.

Proposition 2 The Jones polynomial of the knot $\overset{\square_n}{x_1}$, when n is odd, is

$$V_{\overset{\square_n}{x_1}}(t) = -t^{\frac{1-3n}{2}} + t^{\frac{3-3n}{2}} - t^{\frac{5-3n}{2}} + \cdots + t^{\frac{-n-3}{2}} + t^{\frac{1-n}{2}}.$$

Proof 2 Nothing to prove; just substitute t^{-1} in place of q in $\tilde{\theta}\left(\overset{\square_n}{x_1}\right)$.

REFERENCES

1. E. Witten, "Quantum Field Theory and the Jones Polynomial," Communications in Mathematical Physics, Vol. 121, No. 3, 1989, pp. 351-399.

2. N. Reshetikhin and V. Turaev, "Ribbon Graphs and Their Invariants Derived from Quantum Groups," Communications in Mathematical Physics, Vol. 127, No. 1, 1990, pp. 1-26.

3. V. Turaev, "The Yang-Baxter Equation and Invariants of Links," Inventiones Mathematicae, Vol. 92, No. 3, 1988, pp. 527-553.

4. V. G. Drinfeld, "Hopf Algebras and the Quantum Yang-Baxter Equation," Soviet Mathematics Doklady, Vol. 32, 1985, pp. 254-258.

5. V. G. Drinfeld, "Quantum Groups," Proceedings of the International Congress of Mathematicians (Berkely, 1986)," American Mathematical Society, Providence, 1987, pp. 798-820.

6. M. Jimbo, "A q-Difference Analogue of U(g) and the Yang-Baxter Equation," Letters in Mathematical Physics, Vol. 10, No. 1, 1985, pp. 63-69.

7. V. F. R. Jones, "A Polynomial Invariant for Links via Neumann Algebras," Bulletin of the AMS—American Mathematical Society, Vol. 129, 1985, pp. 103-112.

8. L. H. Kauffman, "State models and Jones Polynomial," Topology Vol. 26, No. 3, 1987, pp. 395-407.

9. V. F. R. Jones, "The Jones Polynomial," Discrete Mathematics, Vol. 294, No. 3, 2005, pp. 275-277.

10. T. Ohtsuki, "Quantum Invariants: A Study of Knots, 3-Manifolds and Their Sets," World Scientific, Sigapore City, 2002.

11. S. Chmutov, S. Duzhin and J. Mostovoy, "Introduction to Vassiliev Knot Invariants," 2011. (A Preliminary Draft Version of a Book on Vassiliev Knot Invariants)

7

CHAPTER

Strong Chang's Conjecture and the tree Property at ω_2

Víctor Torres-Pérez[a,*], Liuzhen Wu[b]

[a] *Institut für Diskrete Mathematik und Geometrie, TU Wien, Wiedner Hauptstraße 8-10/104, 1040 Vienna, Austria*

[b] *Institute of Mathematics, Chinese Academy of Sciences, East Zhong Guan Cun Road No. 55, Beijing 100190, China*

ABSTRACT

We prove that a strong version of Chang's Conjecture together with $2^{\omega}=\omega_2$ implies there are no ω_2-Aronszajn trees.

Keywords: Strong Chang's Conjecture; Tree property

1. INTRODUCTION

Given a regular cardinal κ, we say that κ has the *tree property* if every tree T of height κ and levels of size $<\kappa$, T has a cofinal branch, and it is usually denoted by $\mathrm{TP}(\kappa)$. Trees of height κ with levels of size $<\kappa$ with no cofinal branches are usually called κ-*Aronszajn*.

We list some historical results involving the Tree Property for different regular cardinals. König's Lemma gives some sufficient conditions for a tree to have a cofinal branch. He proved in [5] that $TP(\omega)$ holds. However, Aronszajn showed that we cannot generalize König's Lemma for trees of height ω_1 by constructing an ω_1-Aronszajn tree (see [2]). Considering trees of height ω_2 with levels of size at most $\aleph_1$, it turns out to be independent from the usual axioms of Set Theory. We recall also the result by Silver, where if $TP(\omega_2)$ holds, then $\aleph_2$ is weakly compact in L (Theorem 5.9 in [6]). On the other hand, Mitchell proved that if κ is a weakly compact, then there is a generic extension where $\kappa=\omega_2=2^\omega$ and $TP(\omega_2)$ holds (see [6]). In particular, $TP(\omega_2)$ is equiconsistent with the existence of a weakly compact cardinal.

In these notes we work with a strong version of Chang's Conjecture (see Definition 3.1 of the present notes and also see Theorem 1.3 in [9] for an earlier reference) here denoted by C▯. On one hand, Todorčević and Torres-Pérez proved that under a stronger version of C▯, the negation of CH implies there are no special ω_2-Aronszajn trees (see[15]). On the other hand, Sakai and Velickovic proved that under SSR, a strengthening ofCC▯ (see [1]), the negation of CH together with $MA_{\omega 1}(Cohen)$ implies the strong tree property at ω_2 and so in particular it implies $TP(\omega_2)$ (see [8]).

We prove in these notes that CC▯ and the negation of CH imply $TP(\omega_2)$. Observe that by a result of Todorčević (see [14]), CC▯ implies $2^\omega \le \omega_2$, so under CC▯, ¬CH is equivalent to $2^\omega=\omega_2$.

We make a remark for the necessity of $\neg MA_{\omega 1}(Cohen)$ in [8]:

Theorem 1.1 Folklore.

Assume that there exists a strongly compact cardinal. Then there exists a forcing extension in which $SSR+\neg MA_{\omega 1}(Cohen)+\neg CH$ *holds.*

The following fact is used:

Fact 1.1.

(Shelah [10], Chapter XIII, 1.6 and 1.10) Assume that κ is a strongly compact cardinal. Let $(P_\alpha, \dot{Q}_\beta : \alpha \leq \kappa, \beta < \kappa)$ *be a revised*

countable support iteration of semi-proper posets of size <κ such that $\kappa=\omega_2$ *in* V^{P_κ}. *Then* SSR *holds in* V^{P_κ}.

Assume that κ is strongly compact in V. Let $\langle P_\alpha, \dot{Q}_\beta : \alpha \leq \kappa, \beta < \kappa \rangle$ be the countable support iteration of random forcing. Here recall that a revised countable support iteration coincides with a countable support iteration for proper posets. Note also that $\kappa=\omega_2$ in V^{P_κ}. Hence SSR holds in V^{P_κ} by the above fact. Moreover, $MA_{\omega_1}(\mathrm{Cohen})$ fails in V^{P_κ} as adding random reals makes $\mathbf{non}(B)$ into ω_1.

2. PRELIMINARIES AND BASIC DEFINITIONS

Given a limit ordinal γ, a subset $A\subseteq\gamma$ is *unbounded in* γ if $\sup(A)=\gamma$. A is *closed* in γ if for every limit ordinal $\beta<\gamma$, if $A\cap\beta$ is unbounded in β, then $\beta\in A$. A set $A\subseteq\gamma$ is often called a *club set in* γ if it is closed and unbounded in γ. A set $S\subseteq\gamma$ is stationary, if $S\cap A\neq\emptyset$ for every A club in γ.

The following result involving stationary sets is known as *Fodor's Lemma* or the *Pressing Down Lemma for ordinals.*

Lemma 2.1.

(Fodor [3]) Let κ be a regular uncountable cardinal. Then for every $S\subseteq\kappa$ *stationary, and for every* $f:S\to\kappa$ *such that* $f(\alpha)<\alpha$ *for every* $\alpha\in S$, *there is* $\xi<\kappa$ *such that* $f^{-1}(\{\xi\})$ *is stationary.*

We use a general version of a stationary set, originally by Jech, but in these notes we use an equivalent version due to Kueker (see for example, Theorem 8.28 in [4]). Given an infinite set A, we denote by $[A]^{<\omega}$ the collection of finite subsets of A. Similarly, let $[A]^{\omega}$ denote the collection of all subsets of A of size ω. We say that a set $S\subseteq[A]^{\omega}$ is *stationary in* $[A]^{\omega}$ if for every function $F:[A]^{<\omega}\to A$, there is $X\in[A]^{\omega}$ such that $F(e)\in X$ for every $e\in[X]^{<\omega}$.

The following lemma is the generalized version of the Pressing Down Lemma (see Theorem 8.24 in [4]).

Lemma 2.2 Jech.

For every stationary set $S \subseteq [A]^{\omega}$ *and for every function* $f: S \rightarrow A$ *such that* $f(X) \in X$ *for every* $X \in S$, *there is* $a \in A$ *such that* $f^{-1}(\{a\})$ *is stationary.*

The couple $\langle T, <_T \rangle$ is a *tree* whenever $<_T$ is a partial order of T, and for every $t \in T$, the set $\{s \in T : s <_T t\}$ is well-ordered by $<_T$. Some times we may just write the tree T, assuming there is an implicit order. We denote by $pred_T(t)$ the set of all the $<_T$-predecessors of t in T, and by $ht_T(t) = o.t.(pred_T(t))$. We will denote by $T_{\xi} = \{t \in T : ht_T(t) = \xi\}$. Often we will just drop off the subindex T if the context is clear.

For $A, B \subseteq T$ we denote by $A \perp B$ if for every $s \in A$ and every $t \in B$, s and t are not comparable. Similarly, for $s, t \in T$ and $A \subseteq T$,
let $s \perp t$ and $s \perp A$ iff $\{s\} \perp \{t\}$ and $\{t\} \perp A$ respectively.

Given an ordinal $\lambda \geq \omega_2$, we recall the *Weak Reflection Principle for* λ, $WRP(\lambda)$.

Definition 2.1.

$WRP(\lambda)$ is the following statement: For any stationary subset S of $[\lambda]^{\omega}$, there is $X \subset \lambda$ such that

(1) $|X| = \omega_1$,

(2) $\omega_1 \subseteq X$ and $S \cap [X]^{\omega}$ is a stationary subset of $[X]^{\omega}$.

Todorčević showed the following (see Lemma 6 in [14]):

Lemma 2.3 Todorčević.

CC□ *implies* $WRP(\omega_2)$.

3. MAIN THEOREM

In this section we prove our main result.

Theorem 3.1.

Under $CC^{\square}$, $\neg$CH is equivalent to the tree property at ω_2.

We follow very closely the proof of Theorem 2.2 in [15]. It is a classical result of Specker that $TP(\omega_2)$ implies $\neg$CH (see [11]).

Given two sets $M^{\square}, M$ we will denote by $M^{\square} \sqsupseteq M$ iff $M^{\square} \supseteq M$ and $M^{\square} \cap \omega_1 = M \cap \omega_1$. Consider the following strong version of Chang's Conjecture:

Definition 3.1 $CC^{\square}$.

There are arbitrarily large uncountable regular cardinals θ such that for every well-ordering < of H_θ, and every countable elementary submodel $M \prec \langle H_\theta; \in, < \rangle$, and every ordinal $\eta < \omega_2$, there exists an elementary countable submodel $M^{\square} \prec \langle H_\theta; \in, < \rangle$ such that $M^{\square} \sqsupseteq M$ and $(M^{\square} \cap \omega_2) \setminus \eta \neq \emptyset$.

We will need the following Proposition for the proof of Lemma 3.1, namely in Claim 3.1.

Proposition 3.1.

Let T be a κ-Aronszajn tree (κ a regular cardinal). Given a regular cardinal $\mu < \kappa$, consider a family of collection of nodes $\langle A_\xi : \xi \in X \rangle$ such that X contains a stationary set consisting of ordinals of cofinality at least μ, $A_\xi \subseteq T_\xi$ and $|A_\xi| < \mu$ for every $\xi \in X$. Then for every λ large enough such that $\{\kappa, T, X, \langle A_\xi : \xi \in X \rangle, \ldots\} \subset H_\lambda$ and for every elementary submodel $N \prec \langle H_\lambda; \in, <, \kappa, T, X, \langle A_\xi : \xi \in X \rangle, \ldots \rangle$ such that $A_\xi \subseteq N$ for every $\xi \in X \cap N$, then for every $t \in T$ of height at least $\sup(N \cap \kappa)$ there are unboundedly many (in $\sup(N \cap \kappa)$) $\xi \in X \cap N$ such that every $s \in A_\xi$ is incomparable with t.

Proof.

Suppose otherwise, and take $t \in T$ of height at least $\sup(N \cap \kappa)$ and $\alpha \in N$ such that for all $\xi \in X \cap N \setminus \alpha$, there is a node $t_\xi \in A_\xi$ such that $t_\xi \leq_T t$. Without loss of generality, we can suppose that X is a stationary set consisting of ordinals greater than α and of cofinality at least μ.

Since $|A_\xi|<\mu$ for any $\xi\in X$, there is an ordinal $\beta_\xi<\xi$ such that for any $s,s'\in A_\xi$, $s=s'\leftrightarrow s\restriction\beta_\xi=s'\restriction\beta_\xi$. By elementarity and using Fodor's Lemma, we can find $\beta\in N\cap X$ and a stationary set $S\in N$ such that for any $\xi\in S$, $s=s'\leftrightarrow s\restriction\beta=s'\restriction\beta$ for any $s,s'\in A_\xi$.

Then for every $s\in A_\beta$, we can define a function $f_s:S\to T$ such that $f_s(\xi)$ is the unique $s_\xi\in A_\xi$ such that $s_\xi>s$. Since $A_\beta\subseteq N$, in particular $s=t\restriction\beta\in N$, and therefore f_s is defined in N . However, by our initial assumption, $f_s(\xi)=t_\xi$ for every $\xi\in S\cap N$, and so f_s defines in N a cofinal branch of T, contradiction. □

Let T be an ω_2-Aronszajn tree. In order to simplify the proof, without loss of generality, we suppose that $T\subseteq\omega_2$ and let $e:\omega_2\times\omega_1\to T$ be a bijective function such that $e(\delta,\xi)\in T_\delta$ for every $(\delta,\xi)\in\omega_2\times\omega_1$. Let θ be sufficiently large such that T, e and all relevant parameters are members of H_θ.

Lemma 3.1.

Assume CC* *and that* T *is an* ω_2*-Aronszajn tree. For every* $M\prec H_\theta$ *countable, and for every* $\eta_0,\eta_1\in\omega_2$, *we can find* $M_0,M_1\prec H_\theta$ *countable such that:*

(1) $M\cap\omega_1=M_0\cap\omega_1=M_1\cap\omega_1$,

(2) $M_0\cap\omega_2\setminus\eta_0\neq\emptyset$ *and* $M_1\cap\omega_2\setminus\eta_1\neq\emptyset$,

(3) $\exists\,\delta_0\in(M_0\cap\omega_2)$ *and* $\delta_1\in(M_1\cap\omega_2)$ *such that* $(M_0\cap T_{\delta_0})\perp(M_1\cap T_{\delta_1})$.

Proof.

Fix $\lambda>\theta$ sufficiently large such that CC* holds in H_λ and M,η_0,η_1 and all relevant parameters are in H_λ. Let $N\prec H_\lambda$ such that if $\gamma=\sup(N\cap\omega_2)$, then $cof(\gamma)=\omega_1$.

Fix M_1 witnessing CC* for M and γ.

We need the following Claim:

Claim 3.1.

For every $t\in T$ *of height at least* γ, *there is* $M^*\sqsupseteq M$ *with* $M^*\in N$ *and* $\beta\in M^*\cap\omega_2$ *such that* $t\perp T_\beta\cap M^*$.

Proof.

Assume otherwise, and take $t \in T$ of height at least γ such that for every $M□ \in N$ with $M□ \sqsupseteq M$, for each $\beta \in M□ \cap \omega_2$, there is an $s_\beta \in (T_\beta \cap M□)$ such that $s_\beta < t$.

We work inside N in this paragraph. Using that CC□ holds in N, build a sequence of models $\langle M_\eta : \eta \in \omega_2 \rangle$ such that $M_\eta \sqsupseteq M$ and $M_\eta \cap \omega_2 \setminus \eta \neq \emptyset$ for every $\eta \in \omega_2$. Let β_ξ be the minimum $\beta \in \omega_2 \setminus \xi$ such that there is $\eta \in \omega_2$ such that $\beta_\xi = \min(M_\eta \cap \omega_2 \setminus \eta)$. Let η_ξ be the minimum $\eta \in \omega_2$ such that $\beta_\xi = \min(M_\eta \cap \omega_2 \setminus \eta)$. Define $\langle A_\xi : \xi \in \omega_2 \rangle$ by setting A_ξ to be the set of nodes r in T_ξ with $r \leq s$ for some $s \in M_{\eta_\xi} \cap T_{\beta_\xi}$. Remark that since M_{η_ξ} is countable, so is A_ξ.

By Proposition 3.1, there are unboundedly many $\xi \in N \cap \omega_2$ such that $t \perp A_\xi$, so choose one of such ξ 's. Then there is $s \in M_{\eta_\xi} \cap T_{\beta_\xi}$ such that $s <_T t$. Thus there is $r \in A_\xi$ such that $r \leq_T s <_T t$, contradicting that r and t are incomparable. □

Let $\{t_n : n \in \omega\}$ be an enumeration of $M_1 \cap T \setminus \gamma$. Using Claim 3.1, build a $\subseteq$-increasing sequence $\langle M_n^0 : n \in \omega \rangle$ of countable elementary submodels of H_θ such that for every $n \in \omega$, $M_n^0 \in N$ and $M_n^0 \sqsupseteq M$, and such that there is $\beta \in M_n^0 \cap \omega_2$ with $t_n \perp M_n^0 \cap T_\beta$. Let M_0 be an end-extension of $\bigcup_{n<\omega} M_n^0$ derived from CC□ and η_0. Let $\delta_0 = \min(M_0 \cap \omega_2 \setminus \eta_0)$ and $\delta_1 = \min(M_1 \cap \omega_2 \setminus \gamma)$. We claim it suffices.

Take $s \in T_{\delta_0} \cap M_0$ and $t \in T_{\delta_1} \cap M_1$. In particular, there is $n \in \omega$ and $\beta \in M_n^0 \cap \omega_2$ such that $t = t_n$ and $t \perp T_\beta \cap M_n^0$. Since $\beta \in M_n^0 \subseteq M_0$, we have $s \upharpoonright_\beta \in M_0$. Moreover, since the enumeration function $e \in M_n^0 \subseteq M_0$ and $M_n^0 \cap \omega_1 = M_0 \cap \omega_1$, we have $T_\beta \cap M_0 = T_\beta \cap M_n^0$ and so $s \upharpoonright_\beta \in M_n^0$. Therefore $s \upharpoonright_\beta$ is not comparable with t, and so neither are s and t.

This finishes the proof of Lemma 3.1. □

Lemma 3.2.

Assume CC□. Let T be an ω_2-Aronszajn tree. If the set

$$S_T = \{A \in [\omega_2]^\omega : \forall t \in T(\text{pred}(t) \cap A \textit{ is bounded in } \sup(A))\}$$

is nonstationary, then CH *holds.*

Proof.

Let $f:[\omega_2]^{<\omega} \to \omega_2$ such that the set C_f of closure points of f (i.e. $X \in C_f$ iff for every $e \in [X]^{<\omega}$, $f(e) \in X$) is disjoint with S_T. We can suppose that $T \subseteq \omega_2$ and $e:\omega_1 \times \omega_2 \to T$ is a bijection such that $e(\delta,\beta) \in T_\delta$. Let λ be sufficiently large such that T, S_T, f, e and all relevant parameters are members of H_λ.

Using previous lemma, build a binary tree $\langle M_\sigma \rangle_{\sigma \in 2^{<\omega}}$ of countable elementary submodels of H_λ with the property that for every $\sigma \in 2^{<\omega}$

(1) $M_\sigma \cap \omega_1 = M_{\sigma\frown 0} \cap \omega_1 = M_{\sigma\frown 1} \cap \omega_1$,

(2) $M_\sigma \cap \omega_2 \subsetneq M_{\sigma\frown 0} \cap \omega_2$ and $M_\sigma \cap \omega_2 \subsetneq M_{\sigma\frown 1} \cap \omega_2$,

(3)there exists $\delta_0 \in (M_{\sigma\frown 0} \cap \omega_2)$ and $\delta_1 \in (M_{\sigma\frown 1} \cap \omega_2)$ such that $T_{\delta_0} \cap M_{\sigma\frown 0} \perp T_{\delta_0} \cap M_{\sigma\frown 1}$,

(4)for every $r \in 2^\omega$, if $M_r = \bigcup_{n\in\omega} M_{r\restriction n}$, then for every $r,r' \in 2^\omega$, $\sup(M_r \cap \omega_2) = \sup(M_{r'} \cap \omega_2)$.

Let δ be the common supremum of every $M_r \cap \omega_2$, $r \in 2^\omega$. Then for every $r \in 2^\omega$, there is $t_r \in T_\delta \cap M_r$ such that for every $\text{pred}(t_r) \cap M_r$ is unbounded in δ.

Claim 3.2.

The application $r \mapsto t_r$ *is an injection from* 2^ω *to* T_δ *(and so* CH *does hold).*

Proof.

Let $r_0, r_1 \in 2^\omega$ with $r_0 \neq r_1$ and denote by t_i the node t_{r_i} for $i \in \{0,1\}$. We will find two predecessors of t_0 and t_1 that are incomparable.

Let $n \in \omega$ such that $r_0\restriction_n = r_1\restriction_n = \sigma$, and $r_0\restriction_{n+1} \neq r_1\restriction_{n+1}$. Without loss of generality suppose $r_i(n) = i$ for $i \in \{0,1\}$.

Since $(M_{ri} \cap \omega_2) \notin S_T$, we can find $s_i <_T t_i$ with $s_i \in M_{ri \restriction mi}$ for some $m_i > n$. By the construction of our binary tree, we can take $\delta_0 \in M_{r0 \restriction n+1}$ and $\delta_1 \in M_{r1 \restriction n+1}$ such that $T_{\delta 0} \cap M_{r0 \restriction n+1} \perp T_{\delta 1} \cap M_{r1 \restriction n+1}$. However, observe that for $i \in \{0,1\}, \delta_i \in M_{ri \restriction n+1} \subseteq M_{r1}$, and so $t_i \restriction_{\delta i} \in M_{ri \restriction n+1}$. Therefore, $t_0 \restriction_{\delta 0}$ and $t_1 \restriction_{\delta 1}$ are incomparable, and so $t_0 \neq t_1$. □

This finishes the proof of Lemma 3.2. □

We are now ready to finish the proof of our Theorem. From the previous lemma we know that the set S_T is stationary in $[\omega_2]^{\omega_0}$. Let $S'_T = S_T \cap C_e$, where C_e is the club of all countable subsets of ω_2 closed under the level enumeration function *e* of *T*.

We now use that CC implies $WRP(\omega_2)$ (Lemma 2.3). Take $X \subseteq \omega_2$ of size $\aleph_1$ such that $\omega_1 \subseteq X$ and where $S'_T \cap [X]^\omega$ is stationary. Take $t \in T$ of height at least $\sup(X)$.

From the definition of S_T, for every $A \in S'_T \cap [X]^\omega$ we can choose $\beta_A \in A$ such that if $s \in pred(T) \cap A$, then $s < \beta_A$. By the Pressing Down Lemma, there is a stationary set $S \subseteq S'_T \cap [X]^\omega$ and a β such that $\beta_A = \beta$ for all $A \in S$. Let $\xi \in \omega_1$ such that $e(\beta,\xi) = t \restriction_\beta$. Observe that *S* is in particular cofinal in $[X]^\omega$ so $\bigcup S = X$. Since $\omega_1 \subseteq X$, pick $A \in S$ such that $\xi \in A$. Therefore, $e(\beta,\xi) \in A \cap pred(t)$, and so $e(\beta,\xi) < \beta$. But this is a contradiction, since in general $e(\beta,\xi) \geq \beta$ for any $\beta \in \omega_2$. This ends the proof of our Theorem.

4. SOME FINAL REMARKS

We mention some related previous results. R. Strullu proved that the Map Reflection Principle, introduced by Moore in [7], together with $MA_{\omega 1}$ implies $TP(\omega_2)$ (see [12]). Also it is implicit in B. Velickovic and H. Sakai's results ([8]) that $WRP(\omega_2) + MA_{\omega 1}(Cohen)$ implies $TP(\omega_2)$.

We remark that the results in [15] were in the context of Rado's Conjecture (RC), which is the following statement in Todorčević's equivalent version:

Definition 4.1 RC.

Every tree T of height $\aleph_1$ is special, i.e., the countable union of antichains if and only if every subtree of T of size $\aleph_1$ is also special.

Todorčević proved via a large cardinal that RC is consistent, and showed it is independent from ZFC. In particular, RC is not compatible with $MA_{\omega 1}$ (see final remarks in [13]).

As we have mentioned, in [15], it was proved that Rado's Conjecture together with the negation of the Continuum implies there are no special ω_2-Aronszajn trees. One natural question was which extra condition we could add to Rado's Conjecture to obtain that there are no ω_2-Aronszajn trees at all. Since Rado's Conjecture is consistent with both CH and ¬CH, and CH implies $\neg TP(\omega_2)$, we needed at least to add the condition ¬CH to RC if we wanted to obtain $TP(\omega_2)$. However, as we have mentioned, RC is not consistent with$MA_{\omega 1}$, so we could not have similar results as the one cited above.

Todorčević proved in [14] that RC implies $CC^{?}$. Therefore, a consequence of the result in the present paper is that the condition ¬CH was not only needed, but also sufficient to add to RC to get $TP(\omega_2)$.

Corollary 4.1.

RC *and* ¬CH *imply* $TP(\omega_2)$.

As we have mentioned, Todorčević proved in [14] that $CC^{?}$ implies $WRP(\omega_2)$. The following question is still open.

Question 4.1.

Do $WRP(\omega_2)$ *and* ¬CH *imply together* $TP(\omega_2)$*?*

REFERENCES

1. Philipp Doebler, Ralf Schindler, Π2 consequences of BMM + NSω1 is precipitous and the semiproperness of stationary set preserving forcings, Math. Res. Lett. 16 (5) (2009) 797–815.
2. Kurepa Dura, Ensembles ordonnés et ramifiés, Publ. Math. Univ. Belgr. 4 (1935) 1–38.
3. G. Fodor, Eine Bemerkung zur Theorie der regressiven Funktionen, Acta Sci. Math. (Szeged.) 17 (1956) 139–142.
4. Thomas Jech, Set Theory, Springer Monogr. Math., Springer-Verlag, Berlin, 2003. The third millennium edition, revised and expanded.
5. Dénes Kőnig, Über eine Schlussweise aus dem Endlichen ins Undendliche, Acta Sci. Math. (Szeged.) 3 (2–3) (1927) 121–130.
6. William Mitchell, Aronszajn trees and the independence of the transfer property, Ann. Math. Log. 5 (1972/73) 21–46.
7. Justin Tatch Moore, Set mapping reflection, J. Math. Log. 5 (1) (2005) 87–97.
8. Hiroshi Sakai, Boban Veličković, Stationary reflection principles and two cardinal tree properties, J. Inst. Math. Jussieu 14 (01) (2015) 69–85.
9. Saharon Shelah, Proper Forcing, Lect. Notes Math., vol. 940, Springer-Verlag, Berlin, 1982. [10] Saharon Shelah, Proper and Improper Forcing, second edition, Perspect. Math. Log., Springer-Verlag, Berlin, 1998.
10. E. Specker, Sur un problème de Sikorski, Colloq. Math. 2 (1949) 9–12.
11. Remi Strullu, MRP, tree properties and square principles, J. Symb. Log. 76 (4) (2011) 1441–1452
12. S. Todorčević, On a conjecture of R. Rado, J. Lond. Math. Soc. 27 (1) (1983) 1–8.

13. Stevo Todorčević, Conjectures of Rado and Chang and cardinal arithmetic, in: Finite and Infinite Combinatorics in Sets and Logic, Banff, AB, 1991, in: NATO Adv. Stud. Inst. Ser., Ser. C, Math. Phys. Sci., vol. 411, Kluwer Acad. Publ., Dordrecht, 1993, pp. 385–398.

14. Stevo Todorčević, Víctor Torres-Pérez, Conjectures of Rado and Chang and special Aronszajn trees, Math. Log. Q. 58 (4–5) (2012) 342–347.

8

CHAPTER

Topology of Platonic Spherical Manifolds: From Homotopy to Harmonic Analysis

Peter Kramer

Institut für Theoretische Physik der Universität Tübingen, Tübingen 72076, Germany

ABSTRACT

We carry out the harmonic analysis on four Platonic spherical three-manifolds with different topologies. Starting out from the homotopies (*Everitt 2004*), we convert them into deck operations, acting on the simply connected three-sphere as the cover, and obtain the corresponding variety of deck groups. For each topology, the three-sphere is tiled into copies of a fundamental domain under the corresponding deck group. We employ the point symmetry of each Platonic manifold to construct its fundamental domain as a spherical orbifold. While the three-sphere supports an orthonormal complete basis for harmonic analysis formed by Wigner polynomials, a given spherical orbifold leads to a selection of a specific subbasis. The resulting selection rules find applications in cosmic topology, probed by the cosmic microwave background.

Keywords: topology of Platonic three-manifolds; spherical orbifolds; harmonic analysis on orbifolds

1. INTRODUCTION

In 1917, Einstein replaced the Euclidean three-space by the three-sphere and so introduced the first spherical manifold to describe the spatial part of the Universe [1]. More generally, three-manifolds with non-Euclidean topology have in recent years found applications in cosmology, in particular in relation to the multipole analysis of the cosmic microwave background (CMB) radiation. Cosmological models with positive curvature are related to the three-sphere.

The three-sphere covers all spherical three-manifolds, including the Platonic manifolds considered here. For the Poincarés dodecahedral three-manifold, the first orthonormal basis required for the multipole analysis was constructed in [2], using Lie algebraic methods and drawing on work by Klein 1884 [3].

Compared to other manifolds, the Platonic three-manifolds have the advantage that their homotopies have been derived and classified in 2004 by Everitt [4]. The homotopies are fundamental for the mathematical analysis of the topology of manifolds [5–9]. The connection of multipole-resolved CMB measurements to cosmic topology is provided by selection rules for the harmonic analysis.

Surprisingly, there is no systematic account that links the homotopy of Platonic manifolds to deck groups and to their harmonic analysis and selection rules. Our aim is to close this gap.

We start by relating the three-sphere to the group $SU(2, C)$ in Section 2. The representations of the group $SU(2, C)$ yield the Wigner polynomials introduced in Section 3, with unitary actions resulting in a multipole decomposition discussed in Section 4. In Section 5, we index the Wigner basis polynomials by points on an m-grid in two dimensions in preparation for deriving topological selection rules.

The Platonic spherical polyhedra are constructed in Sections 6 and 7 by use of four spherical Coxeter groups for the tetrahedron, cube, octahedron and dodecahedron. The Platonic homotopy groups from [4] are expressed in Sections 8 and 9 by the gluings of faces and directed edges of the Platonic polyhedra.

In the central Section 10, we use the isomorphism of homotopy and deck groups. Any topological spherical three-manifold appears on its universal cover S^3 in the form of a tiling. Any tile is an image of the spherical manifold. The tiles are related to one another by deck actions from the deck group. We construct from the homotopies of [4] the isomorphic deck actions on the three-sphere S^3 and, from them determine, the deck groups for the Platonic manifolds. The deck groups are factorized in Section 11 by the point symmetry groups of the Platonic polyhedra. We handle fixpoints under rotations in point groups by the new construction of multiply-connected topological orbifolds. Whereas a topological manifold strictly excludes fixpoints under deck actions, orbifolds allow for such fixpoints ([7,9] and pp. 652–714 in [8]).

For the harmonic analysis on orbifolds, we employ in Section 12 Wigner polynomials with restrictions to the subgrid representation of topological selection rules. The resulting subbases appear as denumerable sets of Wigner polynomials on subgrids, as shown in Figure 3 and extend earlier work [10]. The m-grid representation allows in Section 13 for a very transparent graphical display of the restriction from general bases to subbases of a given topology.

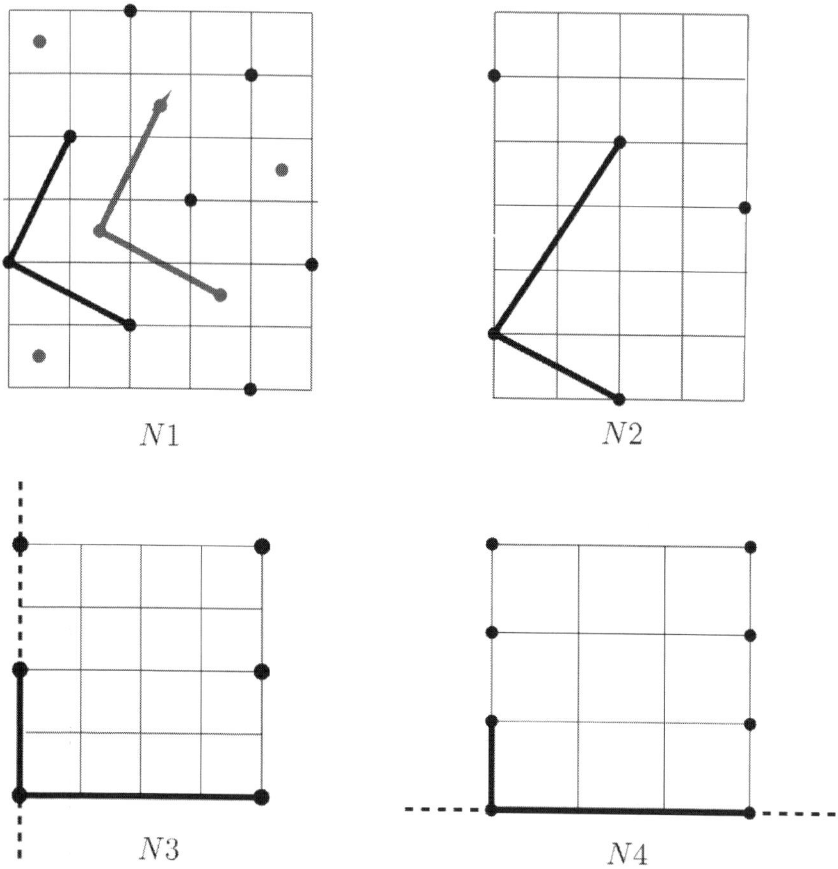

Figure 3. Subgrids of integer/half-integer black/red points $(m_1 m_2)$ are spanned by heavy black/red vectors. The marked grid points select the bases Djm1m2Dm1m2j for a spherical orbifold *Ni* from the square grid formed by thin lines. Coordinate transformations are given in Equation (36). Tetrahedral case *N*1, coordinates u(x˜), integer black and half-integer red grid points. Cubic case *N*2, coordinates u(x˜), with integer black grid points. Cubic case *N*3, coordinates $u(x)$, vertical broken mirror line, integer black grid points. Octahedral case *N*4, coordinates $u(x)$, horizontal broken mirror line, integer black grid points.

Finally in Section 14, we present the application of topology to cosmology. In contrast to Einstein's closure of the spatial part of space-time on the simply-connected three-sphere, we discuss the closure on spherical orbifolds and incorporate a wider variety of multiply-connected cosmologies. Our harmonic analysis on orbifolds yields for each orbifold specific multipole selection rules and, moreover, predicts topological correlations between different multipole orders (l, l') of the CMB.

2. THE THREE-SPHERE S^3 IS UNITARY

In three-dimensional topology, the three-sphere S^3 is the simply-connected universal cover [7] (p. 290) of spherical topologies. The points of the three-sphere, seen in Euclidean space E^4 with covariant coordinates, are:

$$x = (x_0, x_1, x_2, x_3) : \sum_{i=0}^{3} (x_i)^2 = 1 \quad (1)$$

They are in one-to-one correspondence to the elements of the unitary unimodular group $SU(2, C)$. From real coordinates of E^4: $x = (x_0, x_1, x_2, x_3)$, we pass to a 2×2 unimodular unitary matrix $u(x) \in SU(2, C)$ in the form:

$$u(x) = \begin{bmatrix} z_1 & z_2 \\ -\overline{z}_2 & \overline{z}_1 \end{bmatrix} = \begin{bmatrix} x_0 - ix_3 & x_2 - ix_1 \\ -x_2 - ix_1 & x_0 + ix_3 \end{bmatrix}, \ \det(u) = 1 \quad (2)$$

As angular coordinates on $SU(2, C)$, we can use the Euler angles (α, β, γ) [11] (pp. 6–8). In these, the matrix u takes the form:

$$u(\alpha,\beta,\gamma) = \begin{bmatrix} \exp(\frac{i(\alpha+\gamma)}{2})\cos(\frac{\beta}{2}) & \exp(\frac{i(\alpha-\gamma)}{2})\sin(\frac{\beta}{2}) \\ -\exp(\frac{-i(\alpha-\gamma)}{2})\sin(\frac{\beta}{2}) & \exp(\frac{-i(\alpha+\gamma)}{2})\cos(\frac{\beta}{2}) \end{bmatrix} \tag{3}$$

The measure of integration in the Euler angles becomes [11] (p. 62),

$$d\mu(\alpha,\beta,\gamma) = \frac{1}{8\pi^2} d\alpha \sin(\beta) d\beta d\gamma \tag{4}$$

Summary: The three-sphere S^3 corresponds one-to-one to the group *SU*(2, *C*). From this correspondence, the isometries ofS^3 inherit a unitary structure. This provides the full representation theory [11] of the group *SU*(2, *C*) as a tool for spherical topology.

3. WIGNER POLYNOMIALS

The finite irreducible unitary representations of *SU*(2, *C*) were studied by Wigner [12]. For unitarity and the irreducibility of representations, we refer to his monograph. We now introduce the Wigner polynomials. We shall show that they span the harmonic analysis on the three-sphere [2] and play an important part in topological analysis.

Let the matrix $u \in SU(2, C)$ Equation (2) act from the left on the complex column vector:

$$u : \begin{bmatrix} a \\ b \end{bmatrix} \to u(z_1, z_2) \begin{bmatrix} a \\ b \end{bmatrix} \tag{5}$$

In the Bargmann–Hilbert space of analytic functions [13] in two complex variables (*a*, *b*), the monomials:

$$\phi(a,b)^j_m = \frac{1}{\sqrt{(j-m)!(j+m)!}} a^{j+m} b^{j-m}, 2j = 0, 1, 2, ..., -j, \leq m \leq j \quad (6)$$

are normalized with the measure:

$$\mathrm{d}\mu(a,b) = \pi^{-2} \exp(-a\bar{a} - b\bar{b})\, \mathrm{dRe}(a)\mathrm{dIm}(a)\, \mathrm{dRe}(b)\mathrm{dIm}(b) \quad (7)$$

With m either an integer or half-integer, under the action Equation (5), the monomials in Equation (6) carry irreducible representations of *SU*(2, *C*) and transform as:

$$\phi(az_1 + bz_2, -a\bar{z}_2 + b\bar{z}_1)^j_{m_1} = \sum_{m_2} \phi(a,b)^j_{m_2} D^j_{m_2,m_1}(z_1, z_2, \bar{z}_1, \bar{z}_2) \quad (8)$$

The $(2j+1)^2$ coefficients in this equation are the Wigner D^j-functions [11] (Equation (4.1.10)). Wigner [12] has shown that the D^j-functions are the unitary irreducible representations of *SU*(2, *C*). We replace Euler angles by four complex variables from Equation (2) in the arguments of the D^j. Following [2], we term the resulting complex valued polynomials Wigner polynomials. From Equation (8), they are given by:

$$\begin{aligned} D^j_{m_1,m_2}(z_1, z_2, \bar{z}_1, \bar{z}_2) := & \left[\frac{(j+m_1)!(j-m_1)!}{(j+m_2)!(j-m_2)!}\right]^{1/2} \\ & \sum_\sigma \frac{(j+m_2)!(j-m_2)!}{(j+m_1-\sigma)!(m_2-m_1+\sigma)!\sigma!(j-m_2-\sigma)!} \\ & \times(-1)^{m_2-m_1+\sigma} z_1^{j+m_1-\sigma} \bar{z}_2^{m_2-m_1+\sigma} z_2^{\sigma} \bar{z}_1^{j-m_2-\sigma} \end{aligned} \quad (9)$$

All exponents of the four complex variables in Equation (9) are integers and must be non-negative. This restricts the summation over the integer σ. Two important properties of these Wigner polynomials arise under inversion and under complex conjugation of the matrix *u*,

$$D^j_{m_1 m_2}(u^{-1}) = \overline{D^j_{m_2 m_1}(u)},\ D^j_{m_1 m_2}(\overline{u}) = \overline{D^j_{m_1 m_2}(u)} \tag{10}$$

We prove that the Wigner polynomials vanish under the Laplacian on E^4 and, so, are harmonic. Consider first the case $m_1 = j$. This implies $\sigma = j - m_2$,

$$D^j_{j,m_2}(u) = \left[\frac{(2j)!}{(j+m_2)!(j-m_2)!}\right]^{1/2} (z_1)^{j+m_2}(z_2)^{j-m_2} \tag{11}$$

The polynomials Equation (11) are analytic in (z_1, z_2), and so, by the Cauchy–Riemann differential equations, vanish under the Laplacian Δ on E^4. From the particular polynomials Equation (11), we can lower the eigenvalue $m_2 = j$ to any value $m_2 = j - 1, \ldots, -l$ for fixed degree $2j$ by the application of the left lowering operator Ll–L–l of $SU(2, C)^C$, given from [2] (Equation (35)) by:

$$L^l_- = \left[\overline{z}_2 \partial_{z_1} - \overline{z}_1 \partial_{z_2}\right] \tag{12}$$

By repeated application of this lowering operator, we can reach all of the $(2j+1)^2$ Wigner polynomials for fixed degree $2j$. The lowering operator commutes with the Laplacian Δ on E^4 and, so, cannot change its eigenvalue zero. It follows that all of the Wigner polynomials Equation (9) are harmonic.

The action of the rotation group $SO(4;R)$ is expressed by left and right actions on $SU(2;C)$. When restricted to conjugation, its action reduces to rotations only of the three coordinates $(x_1, x_2, x_3) \in E^3$; see Equation (24). The orthogonality of the Wigner polynomials, expressed in the Euler angles, is [11] (p. 62)

$$\frac{1}{8\pi^2}\int \overline{D^{j'}_{m_1' m_2'}(\alpha\beta\gamma)} D^{j}_{m_1 m_2}(\alpha\beta\gamma) d\alpha \sin(\beta) d\beta d\gamma$$
$$= \delta_{j',j}\delta_{m_1',m_1}\delta_{m_2',m_2}\frac{1}{2j+1} \quad (13)$$

Summary: The Wigner polynomials, Equation (6), are harmonic, orthonormal and homogeneous of degree $2j = 0, 1, 2, \ldots, \infty$. Their transformation properties under $SU(2, C)$ are well known from the theory of angular momentum in quantum mechanics [11].

4. UNITARY ACTIONS AND REPRESENTATIONS

The group of isometries of S^3 is $SO(4, R)$. The action of this group is isomorphic to the direct product of a left and a right group $SU^l(2, C)$, acting on $u \in SU(2, C)$ in the form:

$$SO(4, R) \sim (SU^l(2, C) \times SU^r(2, C))/Z_2 \quad (14)$$

The left and right action groups in Equation (14) commute and act as:

$$(g_l, g_r) \in (SU^l(2, C) \times SU^r(2, C)) : u \to g_l^{-1} u g_r \quad (15)$$

The subgroup Z_2 in Equation (14) is generated by $(g_l, g_r) = (-e, -e) \in (SU^l(2, C)$ The direct product formEquation (14), contrary to what is asserted in [14] (p. 277), in general does not extend to subgroups of $SO(4, R)$ and, so, cannot yield their classification. A counter-example is provided by the deck group C_5 of the tetrahedral manifold (Equation (42)), which entangles left and right actions.

Among the actions described by Equation (15) are the conjugation actions:

$$(g_l, g_l) : u \to g_l^{-1} u g_l \tag{16}$$

which form a subgroup $SU(2, C)^C$. The action Equation (16) preserves the trace,

$$\text{trace}(u(x)) = 2x_0 \tag{17}$$

and therefore operates only on the subspace with coordinates $(x_1, x_2, x_3) \in E^3$.

The unitary structure of the three-sphere governs its isometries and introduces the representation theory of $SU(2, C)$.

In particular, we obtain for the action of $SO(4, R)$ on a Wigner polynomial Equation (9):

$$(T_{(g_l,g_r)} D^j_{m_1,m_2})(u) = D^j_{m_1,m_2}(g_l^{-1} u g_r) = \sum_{m_1' m_2'} D^j_{m_1' m_2'}(u)\, D^{(j,j)}_{(m_1' m_2', m_1 m_2)}(g_l, g_r) \tag{18}$$

$$D^{(j,j)}_{(m_1' m_2', m_1 m_2)}(g_l, g_r) = D^j_{m_1 m_1'}(g_l^{-1}) D^j_{m_2' m_2}(g_r) \tag{19}$$

The expression in the last line is the matrix element of the irreducible representation $D^{(j,j)}(g_l, g_r)$ of $SO(4, R)$ for the pair $(g_l, g_r) \in SU^l(2, C) \times SU^r(2, C)$. The degree $2j$ stays fixed under the action of $SO(4, R)$, while the pairs (m_1, m_2) take $(2j + 1)^2$ values.

The actions by conjugation Equation (16) of the subgroup $SU(2, C)^C$ reduce into irreducible form upon transforming the Wigner polynomial basis; compare also [15], with Wigner coefficients of $SU(2, C)$ [11] (pp. 31–52), into a spherical basis by:

$$\begin{aligned} \psi_{\beta lm}(u) &= \delta_{\beta,2j+1} \sum_{m_1 m_2} D^j_{m_1,m_2}(u) C(jm_1m_2|lm) \\ C(jm_1m_2|lm) &:= \langle j - m_1 j m_2 | lm \rangle (-1)^{(j-m_1)}, \quad l = 0, 1, \ldots, 2j, \quad 2j = \beta - 1 \end{aligned} \tag{20}$$

Equation (20) implies that the irreducible representation $D^{(j,j)}$ of $SU(4, R)$ when reduced to the subgroup $SU(2, C)^C$contains any irreducible representation D^l of $SU(2, C)^C$ once and only once for l = 0, ..., 2j. The transformation inverse toEquation (20) is:

$$D^j_{m_1,m_2}(u) = \delta_{\beta,2j+1} \delta_{m,-m_1+m_2} \sum_{l=0}^{2j} \psi_{\beta lm}(u) C(jm_1m_2|lm), \quad l \in Z, \ -l \leq m \leq l \tag{21}$$

The two transformations Equations (20) and (21) will play a crucial role in the recursive construction of the harmonic analysis for orbifolds in Section 12.4. In cosmic topology, Section 14, the spherical basis Equation (20) in terms of spherical harmonics YlmYml is required for the multipole analysis of physical observables, like the CMB radiation. It is given by:

$$\psi_{\beta lm}(u) = R_{\beta l}(\chi) Y^l_m(\theta, \phi) \tag{22}$$

$$R_{\beta l}(\chi) = 2^{l+1/2} l! \sqrt{\frac{\beta(\beta - l - 1)!}{\pi(\beta + l)}} C^{l+1}_{\beta - l - 1}(\cos(\chi)) \tag{23}$$

where Cl+1β−l−1 denotes the Gegenbauer polynomial.

As a result of Equation (20), the conjugation action of $SU(2, C)^C$ on the spherical basis becomes:

$$(T_{(g,g)}\psi)_{\beta lm}(u) = \psi_{\beta lm}(g^{-1}ug) = \sum_{m'=-l}^{l} \psi_{\beta lm'} D^l_{m',m}(g) \tag{24}$$

identical to the irreducible action of the rotation operator T_g on the usual spherical harmonics YlmYml in E^3 [11] (pp. 53–67). The coordinate x_0 in E^4 is unchanged under this action, and so, we find:

Summary: The action of the rotation $(g_l, g_r) \in SO(4, R)$ on Wigner polynomials is given by the representation $D^{(j,j)}(g_l, g_r)$,Equation (18). The rotations $(g, g) \in SO(3, R)$ act only on the first three coordinates, and the spherical basis Equation (20)yields the decomposition of Wigner polynomials into irreducible subbases characterized by the multipole order l.

5. DISCRETE 2*D* *M*-GRIDS

To display the topological selection rules for Wigner polynomials, we concentrate on their discrete labels (m_1, m_2). Both of them are an integer or half-integer. These labels form an integer plus a half-integer 2*D* *m*-grid, both of spacing $(\pm1, \pm1)$, on a 2*D* plane. From the point of view of representations of $SU(2, C)$, the grid points fix subrepresentations of left and right subgroups $U(1)$. Consider the set of Wigner polynomials as being attached to the grid points (m_1, m_2). At a given grid point (m_1, m_2), the labels j of all Wigner polynomials attached to it have the denumerable range $j = j_0 \ldots, \infty$, $j_0 = \mathrm{Max}(|m_1|, |m_2|)$. Conversely, the Wigner polynomials for fixed degree $2j$ occur with values (m_1, m_2) on a centered square $(|m_1|, |m_2| \leq j)$. In Section 12.2, we shall see that harmonic analysis on topological orbifolds selects Wigner bases on subgrids.

Summary: A denumerable set of Wigner polynomials $D^j(u)$ with the index pair (m_1, m_2) and degree $2j$, $j \geq (|m_1|, |m_2|)$ is associated with any single point (m_1, m_2) of a 2*D* *m*-grid.

6. SPHERICAL COXETER GROUPS FOR THE PLATONIC POLYHEDRA

In this section, we turn from the three-sphere to the spherical manifolds of Platonic polyhedra shown in Figure 1. To describe their geometrical transformation properties, we use spherical Coxeter groups. Coxeter groups Γ are generated by Weyl reflections in (hyper-)planes of Euclidean space E^n. Any Weyl reflection W_a is characterized by a unit vector a, normal to the reflection (hyper-)plane. The action of the Weyl reflection W_a on $x \in E^n$ is given by:

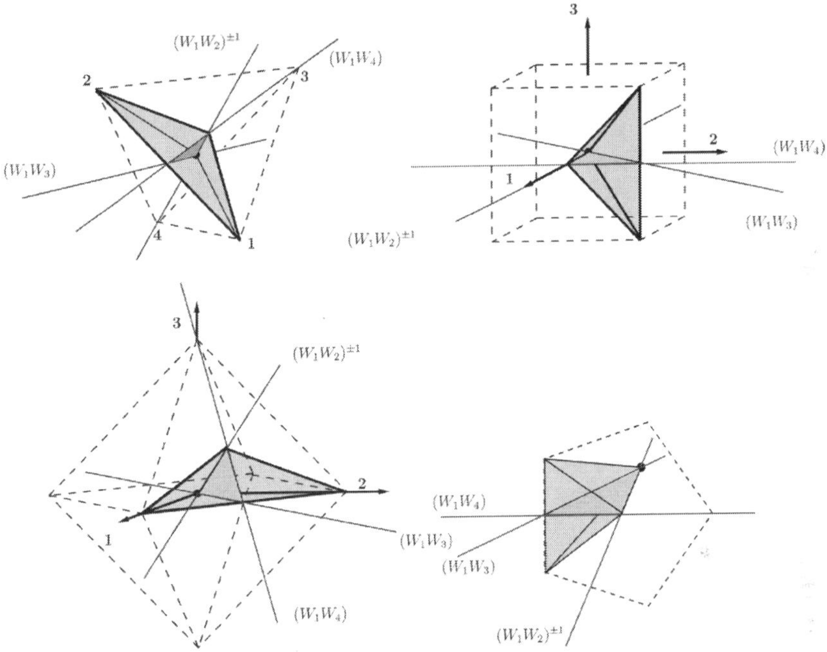

Figure 1. Geometric orbifolds $N1$, $N2$ are shown in the top panels and $N3$, $N9$ in the bottom panels. The yellow color marks the duplex form of the orbifolds as pairs of Coxeter simplices, glued with triangular faces in blue color. The axes of the covering rotations (W_aW_b) for each orbifold are marked by thin lines. The parts of these axes that bound the glue triangles carry fixpoints of integer order p.

$$W_a : x \to W_a x = x - 2\langle x, a \rangle,\ (W_a)^2 = I \tag{25}$$

A Weyl reflection is a Euclidean isometry of determinant −1. A Coxeter group Γ is generated by Weyl reflections W_{ai}, $i = 1, 2, \ldots$ For a fixed Coxeter group Γ, we use the short-hand notation $W_{a1} = W_i$. Γ is graphically denoted by a Coxeter diagram, built from circles for each generator and connected by a sequence of lines. The diagram allows one to construct the group Γ from its Weyl reflection generators and their relations (pp. 12–23 in [16] and [17]). The lines between circles carry integer numbers m_{ij}. Lines without numbers are short-hand for $m_{ij} = 3$. The number $m_{i,i+1}$ between two successive reflections W_{ai}, W_{ai+1} implies the relation:

$$(W_{a_i} W_{a_{i+1}})^{m_{i,i+1}} = I \qquad (26)$$

The number $m_{i,i+1}$ is the order of the rotation generated by $W_{ai}W_{ai+1}$.

In Tables 1 and 2, we list the four spherical Coxeter groups of the tetrahedron, cube, octahedron and dodecahedron, each with four generators, their diagrams and quadruples of unit vectors a_1, ..., a_4. These Coxeter groups produce tilings [18] of the three-sphere S^3 by Platonic polyhedra.

Table 1. Diagrams of four spherical Coxeter groups Γ of order |Γ|, four Platonic polyhedra M and seven deck groups H = deck (M) of order $|H|$ according to [19] (p. 134). In the table, C_n denotes a cyclic, Q the quaternion, T2T2 the binary tetrahedral and J2J2 the binary icosahedral group. The symbols Ni are adapted from [4].

Coxeter Diagram Γ	|Γ|	Polyhedron $\mathcal{M}$	$H = \text{deck}(\mathcal{M})$	|H|	Reference
o – o – o – o	120	tetrahedron $N1$	C_5	5	[20]
o $\overset{4}{–}$ o – o – o	384	cube $N2$	C_8	8	[21]
		cube $N3$	Q	8	[21]
o – o $\overset{4}{–}$ o – o	1152	octahedron $N4$	$C_3 \times Q$	24	[10]
		octahedron $N5$	B	24	[10]
		octahedron $N6$	$\mathcal{T}_2$	24	[10]
o – o – o $\overset{5}{–}$ o	$120 \cdot 120$	dodecahedron $N9$	$\mathcal{J}_2$	120	[2]

Table 2. The four Weyl vectors a_s, $s = 1,\ldots, 4 \in E^4$ for the four Coxeter groups Γ listed inTable 1, with τ:=1+5√2.

Γ	a_1	a_2	a_3	a_4
$\circ - \circ - \circ - \circ$	$(0,0,0,1)$	$(0,0,\sqrt{\frac{3}{4}},\frac{1}{2})$	$(0,\sqrt{\frac{2}{3}},\sqrt{\frac{1}{3}},0)$	$(\sqrt{\frac{5}{8}},\sqrt{\frac{3}{8}},0,0)$
$\circ \overset{4}{-} \circ - \circ - \circ$	$(0,0,0,1)$	$(0,0,-\sqrt{\frac{1}{2}},\sqrt{\frac{1}{2}})$	$(0,\sqrt{\frac{1}{2}},-\sqrt{\frac{1}{2}},0)$	$(-\sqrt{\frac{1}{2}},\sqrt{\frac{1}{2}},0,0)$
$\circ - \circ \overset{4}{-} \circ - \circ$	$(0,\sqrt{\frac{1}{2}},-\sqrt{\frac{1}{2}},0)$	$(0,0,-\sqrt{\frac{1}{2}},\sqrt{\frac{1}{2}})$	$(0,0,0,1)$	$(\frac{1}{2},\frac{1}{2},\frac{1}{2},\frac{1}{2})$
$\circ - \circ - \circ \overset{5}{-} \circ$	$(0,0,1,0)$	$(0,-\frac{\sqrt{-\tau+3}}{2},\frac{\tau}{2},0)$	$(0,-\sqrt{\frac{\tau+2}{5}},0,-\sqrt{\frac{-\tau+3}{5}})$	$(\frac{\sqrt{2-\tau}}{2},0,0,-\frac{\sqrt{\tau+2}}{2})$

A spherical Coxeter group possesses a fundamental domain. On E^4, this is a simplex bounded by the four Weyl reflection planes from the generators of Γ. For the Platonic manifolds, duplex pairs of them are shown in Figure 1.

Summary: Each Platonic spherical polyhedron is linked to one of the four Coxeter groups listed in Tables 1 and 2.

7. THE UNIMODULAR SUBGROUPS *S*Γ

Any element of a Coxeter group from Tables 1 and 2 containing an even number of reflections is a rotation. For topology, we need these rotations, since they preserve orientation. Their 4×4 representations on E^4 are unimodular with determinant one. The subgroup of a Coxeter group Γ generated by all rotations has unimodular matrix representations, and we denote it by $S\Gamma$. This subgroup for all four Coxeter groups in Table 1 is generated in each case by the three products $((W_1W_2), (W_2W_3), (W_3W_4))$ of Weyl reflections.

We wish to pass from a product (W_aW_b) of two Weyl reflection operators to the standard form Equation (18) of $SO(4, R)$ with two $SU(2, C)$ group parameters (g_l, g_r). First, we determine from the unit reflection vectors a, b two unitary matrices$v(a)$, $v(b)$ by use of Equation (2),

$$v(a) := \begin{bmatrix} a_0 - ia_3 & a_2 - ia_1 \\ -a_2 - ia_1 & a_0 + ia_3 \end{bmatrix} \tag{27}$$

and similarly for $v(b)$. Then, as shown in [22] (Equation (11)), we find between products of Weyl operators and $SO(4, R)$ actions Equation (18) the operator relation:

$$T_{W_a W_b} \equiv T_{(g_l, g_r)} : (g_l, g_r) = (v(a)v^{-1}(b), v^{-1}(a)v(b)) \tag{28}$$

Summary: The unimodular subgroup $S\Gamma$ of a Coxeter group Γ is generated by products $(W_a W_b)$, $a \neq b$ of pairs of Weyl reflection operators. Any such product can be converted by Equations (27) and (28) into a rotation operator with parameters (g_l, g_r).

8. HOMOTOPY WITH SPHERICAL POLYHEDRA

There are five spherical Platonic polyhedra. Everitt in [4] applied Sim's low index subgroups algorithm to determine all possible homotopies of these polyhedra. His results are given in the form of diagrams for numbered faces and edges and identification of pairs of faces and edges. The identification is also referred to as gluings and by construction encodes the first homotopy or fundamental group.

We prefer to list the gluings of faces and edges, enumerated in line with Everitt [4] and shown in Figure 2. Note that a given spherical polyhedron may have several different inequivalent homotopy groups. As standard examples, we take the cubic spherical manifolds [21]. This manifold admits two inequivalent homotopies, which we denote as $N2$, $N3$. We depict the polyhedra in an Euclidean fashion. For the cube, an enumeration of the six faces $F\ i$, $i = 1, \ldots, 6$ and twelve directed edges 1, ..., 12 following Everitt [4] is shown in Figure 2. Any square face is characterized by a directed sequence of four edges. Any homotopic gluings of faces plus edges is a map between two such sequences. As an example, we illustrate a gluing $F1 \cup F3$ in $N2$ as a map from right to left,

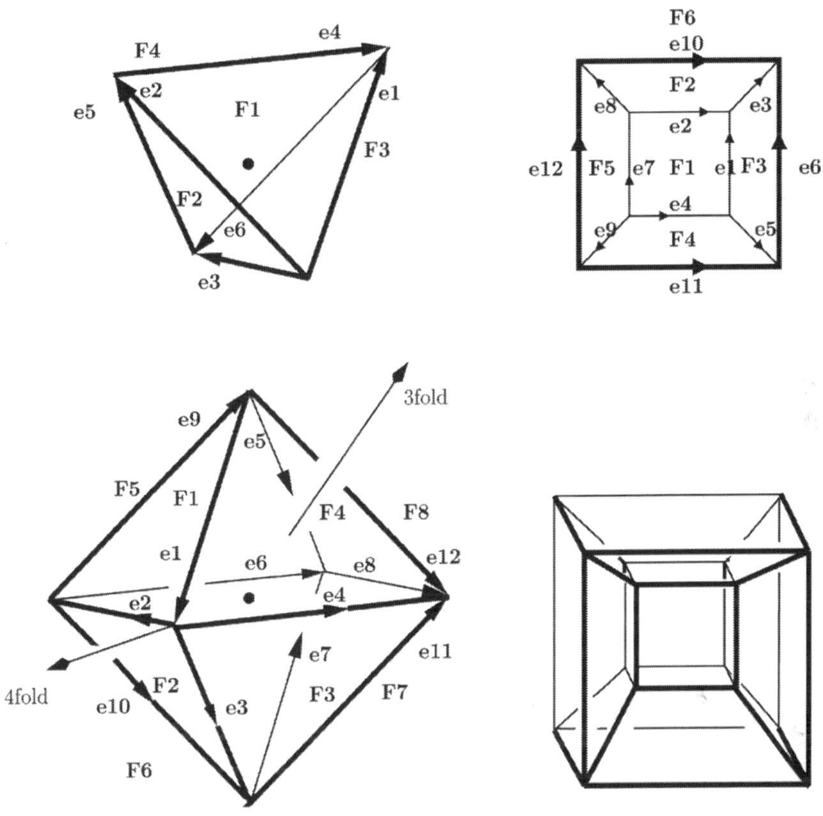

Figure 2. Face and edge enumeration of the tetrahedron, cube, octahedron and the eight-cell that divides the three-sphere into eight spherical cubes.

$$g_1(1 \Leftarrow 3) = \begin{bmatrix} & \bar{1} & \\ \bar{4} & & 2 \\ & 7 & \end{bmatrix} \Leftarrow \begin{bmatrix} & \bar{3} & \\ \bar{1} & & 6 \\ & 5 & \end{bmatrix} \quad (29)$$

From [4], we find for the first cubic homotopy group $N2$ the pairwise face gluings:

$$F3 \cup F1, F4 \cup F2, F6 \cup F5 \tag{30}$$

The glued directed edges are listed on single lines in the diagram given in Table 3.

Table 3. The homotopies of Platonic spherical polyhedra in terms of face and edge gluings. The directed edges in each line of the last column are glued.

Polyhedron	Face Glue	Edge Glue
Tetrahedron $N1$	$F3 \cup F1,\ F2 \cup F4$	$\begin{bmatrix} 1 & \overline{3} & \overline{4} \\ 2 & \overline{5} & \overline{6} \end{bmatrix}$
Cube $N2$	$F3 \cup F1,\ F4 \cup F2,\ F6 \cup F3$	$\begin{bmatrix} 1 & 3 & 4 \\ 2 & 6 & \overline{9} \\ 5 & 7 & \overline{10} \\ 8 & 11 & \overline{12} \end{bmatrix}$
Cube $N3$	$F1 \cup F6, F2 \cup F4, F3 \cup F5$	$\begin{bmatrix} 1 & 8 & 11 \\ 2 & \overline{6} & \overline{9} \\ 3 & 4 & \overline{12} \\ 5 & \overline{7} & \overline{10} \end{bmatrix}$
Octahedron $N4$	$F6 \cup F2, F5 \cup F3, F1 \cup F4, F7 \cup F8$	$\begin{bmatrix} 1 & 4 & 9 \\ 2 & 7 & \overline{12} \\ 3 & 6 & \overline{10} \\ 5 & 8 & 11 \end{bmatrix}$
Octahedron $N5$	$F6 \cup F8, F1 \cup F4, F2 \cup F7, F3 \cup F5$	$\begin{bmatrix} 1 & 4 & 9 \\ 2 & \overline{7} & \overline{12} \\ 3 & 6 & 8 \\ 5 & \overline{10} & 11 \end{bmatrix}$
Octahedron $N6$	$F6 \cup F4, F5 \cup F3, F8 \cup F2, F7 \cup F1$	$\begin{bmatrix} 1 & 8 & 10 \\ 2 & 5 & 11 \\ 3 & 6 & 12 \\ 4 & 7 & 9 \end{bmatrix}$

Summary: All possible homotopies for the Platonic spherical polyhedra are determined and listed in Table 3 in accord with [4].

9. SYNOPSIS OF PLATONIC HOMOTOPIES

The enumeration of faces and directed edges of the Platonic polyhedra is shown in Figure 2. In Table 3, we list the gluing of faces and directed edges as given in [4].

10. FROM HOMOTOPIES TO DECK ACTIONS ON S^3

Homotopic gluings relate faces and edges of a single polyhedron. A new geometric view of topologies emerges on the universal covering manifold S^3. A general theorem, given by Seifert and Threlfall [5] (pp. 181–203), proves for topological manifolds the isomorphism between the first homotopy group and the group of deck or covering actions on the universal cover.

We implement this theorem for the cubic spherical manifold $N2$ and convert its homotopic gluings from Table 3 into deck actions between neighboring copies of a single proto-cube with fixed face and edge enumeration. The deck rotations, which generate the tiling of S^3, form the topological deck group. The eight cubic copies tile the three-sphere in the form of the eight-cell shown in Figure 2 in a projection from [18] (p. 170). This means that the topological deck group must be of order eight.

For the homotopy of $N2$, we make full use of the prescriptions of Table 3. Any quadratic face of the proto-cube is surrounded by four directed edges given in Figure 2. These four directed edges in turn determine the face and, moreover, give its orientation with respect to a four-fold axis perpendicular to the face.

We use square diagrams as given in Equation (31), formed from four directed edges, to denote edges around a face plus their orientation. This orientation must be respected in the homotopic gluing of faces and their edges given in Table 3.

It suffices to convert only the homotopic face gluings of a proto-cube, the glue generators, into deck actions. Then, these deck actions by multiplication generate the full deck group.

In E^3, we choose the coordinates (x_1, x_2, x_3) normal to the three faces labeled $F1$, $F3$, $F2$ in Figure 2. Positive rotations by an angle ϕ around these normals we denote as $R_i(\phi)$, $i = 1, 2, 3$.

Before turning to the manifold $N2$ in Equation (32), we first treat in Equation (31) a simple reference glue $F1 \cup F6$ of two opposite faces of the proto-cube. We convert this glue into a deck action of two copies of the proto-cube and write it algebraically in Equation (31) as a product of Weyl reflections and inversions. By J3J3, J4J4, we denote the inversions in E^3,E^4. We find this action in the form:

$$st(1 \Leftarrow 6) = \begin{bmatrix} & \overline{4} & \\ 7 & & \overline{1} \\ & 2 & \end{bmatrix} \Leftarrow \begin{bmatrix} & \overline{10} & \\ \overline{6} & & 12 \\ & \overline{11} & \end{bmatrix}$$
$$= W_4 \mathcal{J}_3 = W_4 W_0 \mathcal{J}_4 \tag{31}$$

The product of Weyl reflections and inversions in Equation (31) combines into an overall unimodular rotation matrix. For convenience, we introduced the additional Weyl reflection operator W_0, $a_0 = (0, 0, 0, 1)$.

Now, we return to the cubic manifold $N2$. By use of two rotations, we convert its first face gluing, $st(1 \Leftarrow 3) : F1 \cup F3$, into the reference deck action Equation (31),

$$st(1 \Leftarrow 3) = R_1(\pi/2) st(1 \Leftarrow 6) R_3(\pi/2) \tag{32}$$

By similar conjugations as in Equation (31), we convert the two other face gluings of the manifold $N2$ in Equation (30) into deck actions,

$$\begin{aligned} st(2 \Leftarrow 4) &= (W_3 W_2) st(1 \Leftarrow 6)(W_2 W_3) \\ st(3 \Leftarrow 5) &= (W_2 W_3) st(1 \Leftarrow 6)(W_3 W_2) \end{aligned} \tag{33}$$

From products of two Weyl reflection operators in Equation (33), we pass to elements $(g_l, g_r) \in SO(4, R)$ given in Table 2 byEquation (28). As shown in [22], the deck generator in Equation (32) generates a cyclic group C_8 of order eight, which paves the eight-cell tiling of S^3 by copies of the prototile.

The eight-cell admits a second inequivalent cubic homotopy $N3$ [4]. Its deck group derived in [22] is the quaternion group$Q$. We give the generating elements in Equation (48).

For the other Platonic homotopies studied by Everitt [4], the conversions to deck actions are carried out in [20–23].

Summary: The homotopic gluings for Platonic polyhedra from [4] are listed in Table 3. We convert them into actions of deck groups H on the covering three-sphere. They generate tilings by $|H|$ tiles of the covering three-sphere.

11. FROM POINT SYMMETRY TO ORBIFOLDS

The spherical Platonic polyhedra cover the three-sphere by deck transformations. By use of its point symmetry group M, we can decompose a single Platonic polyhedron under M into a fundamental domain and its orbit. We must remove two obstacles towards a topological interpretation:

i. To keep the orientation under deck actions, we restrict the point group M to proper rotations. For the Coxeter group action, its fundamental domain is a simplex bounded by four Weyl reflection planes. When we restrict to rotations to preserve orientation, the actions belong to the unimodular subgroup of Γ of determinant one, which we denote as $S\Gamma$. When passing to $S\Gamma$, we must extend the fundamental simplex domain of Γ for $S\Gamma$ to a duplex, consisting of a simplex and its image under a reflection, see Figure 1.

ii. A topological manifold strictly excludes fixpoints under deck actions. Here, they appear on the rotation axes of the point groups. We must extend our topological notion from a manifold to an orbifold [7,9], which allows for fixpoints of finite order. The notion of orbifolds is explained in more detail in [8] (pp. 652–714). In Figure 1, we give the construction of orbifolds by gluing pairs of Coxeter simplices into duplices for the four Platonic polyhedra. The fixpoints appear on the edges of the blue glue triangles for the two Coxeter simplices.

To cover the full three-sphere by orbifolds, we proceed in two steps. First of all, any Platonic polyhedron under its rotational point symmetry group M is paved by $|M|$ copies of spherical orbifolds. These spherical orbifolds we take as duplices glued from a Coxeter simplex and its mirror image. Copies of the duplices shown in Figure 1 under the point group M tile the Platonic polyhedra, but have fixpoints on their edges. To cover the full three-sphere, we augment the rotational point symmetry group M by the operations of the full deck group H of the Platonic manifold. These two groups share only the identity element, $M \cap H = e$. The product $M \cdot H$ of point and deck operations $m \in M, h \in H$, acting on a proto-orbifold, generates as the deck group the unimodular subgroup $M \cdot H = S\Gamma$ of the Coxeter group. The images of the proto-duplex under all of these products cover the three-sphere.

Summary: The topological orbifolds for Platonic spherical polyhedra are duplices (Figure 1), glued from two Coxeter simplices. Their deck groups are the unimodular subgroups $S\Gamma \in \Gamma$. Each orbifold deck group is factorized as $S\Gamma = M \cdot H$, $M \cap H = e$, into the point symmetry group M and the deck group H for the Platonic polyhedron.

12. HARMONIC ANALYSIS FOR ORBIFOLDS

As the general basis for all of these topologies, we take the harmonic Wigner polynomials. To adapt them to specific orbifolds, we shall present in the following subsections different routes. By graphical means, we shall illuminate on $2D$ m-subgrids; see Figure 3, denumerable sets of Wigner polynomials for orbifolds, selected from the full harmonic set on the three-sphere. More details of the harmonic analysis on the Platonic orbifolds are given in Section 13.

12.1. Rotation to the Diagonal Form

The unitary structure of the three-sphere allows one to reduce any given deck action to a pure phase transformation. A general rotation operator $T_{(gl,gr)}$, $(g_l, g_r) \in SO(4, R)$ can be brought to a simple form by unitary diagonalizations of the matrices (g_l, g_r),

$$g_l = q\delta_l q^{-1},\ g_r = k\delta_r k^{-1}, \quad q, k \in SU(2, C),$$
$$\delta_l = \begin{bmatrix} \exp(i\alpha) & 0 \\ 0 & \exp(-i\alpha) \end{bmatrix}, \delta_r = \begin{bmatrix} \exp(i\gamma) & 0 \\ 0 & \exp(-i\gamma) \end{bmatrix} \tag{34}$$

To find the angular parameters α, γ of the diagonal form, one can avoid the explicit diagonalization Equation (34). For any$u \in SU(2, C)$, use the trace relation:

$$\text{trace}(g_l) = 2\cos(\alpha),\ \text{trace}(g_r) = 2\cos(\gamma) \tag{35}$$

to determine (α, γ).

We now pass from $u(x)$ to a new matrix $u(\tilde{x})$ by the $SO(4, R)$ rotation:

$$q, k: \; u(x) \rightarrow u(\tilde{x}) = q^{-1} u(x) k \tag{36}$$

The matrix u(x˜) represents a new system of coordinates on S^3. It follows from Equation (34) that to the action of $T_{(gl,gr)}$on the coordinates $u(x)$, there corresponds the action $T_{(\delta l,\delta r)}$ on the coordinates u(x˜) If we pass in a Wigner polynomial from the coordinates x˜x˜ to Euler angles, we find in this parametrization:

$$(T_{(\delta_l,\delta_r)} D^j_{m_1,m_2})(u(\tilde{x})) = D^j_{m_1,m_2}(u(\tilde{x})) \, \exp(i(-m_1\alpha + m_2\gamma)) \tag{37}$$

that is, a pure phase transformation. The condition of invariance of a Wigner polynomial Djm1,m2(u′) under the diagonal action of the operator $T_{(\delta l,\delta r)}$ now takes the form of a phase condition:

$$-m_1\alpha + m_2\gamma = \nu\, 2\pi, \; \nu = 0, \pm 1, \pm 2, ... \tag{38}$$

Summary: Any deck action on Wigner polynomials by a rotation can be converted into a pure phase transformationEquation (37). Any deck group with a single-generator can be treated in this way.

12.2. Selection on m-Subgrids

The harmonic analysis must be restricted to the modes admissible for a specific spherical topology. A denumerable set of Wigner polynomials $D^j(u)$ with the index pair (m_1, m_2) and degree $2j$, $j \geq (|m_1|, |m_2|)$ is achieved by choosing from the harmonic Wigner basis a subbasis invariant under the relevant point and deck group. Since all deck actions are of finite order, the angles α, γ in Equation (37) must be integral fractions of 2π of the forms α=p12πq,γ=p22πq For groups with a single generator, rewrite Equation (38) in the form:

$$\begin{aligned} &-m_1\alpha + m_2\gamma = \nu\, 2\pi, \; \alpha = p_1\frac{2\pi}{q}, \gamma = p_2\frac{2\pi}{q}, \\ &-m_1 p_1 + m_2 p_2 = \nu\, q, \; \nu = 0, \pm 1, \pm 2, \ldots \end{aligned} \tag{39}$$

The solutions $(m_1, m_2) = (p_2, p_1)$ of Equation (39) for $\nu = 0$ form integer multiples of a grid vector $a_1 = (p_2, p_1)$. Solutions ofEquation (39) for $\nu \neq 0$ are on lines parallel to a_1. We choose a second vector $a_2 = (q_1, q_2)$ to a grid point outside, but next to the line spanned by a_1. Then, the vectors (a_1, a_2) form the basis of an *m*-subgrid, whose points yield all solutions of Equations (38) and (39). This case applies to the orbifolds *N*1 and *N*2 with cyclic deck groups C_5, C_8. In Figure 3, we mark the subgrids for the orbifolds *N*1, *N*2, *N*3, *N*4.

Note that the topological rules Equation (39), given in terms of subgrids, select enumerable sets of harmonic Widner basis polynomials of unrestricted degree 2*j*.

12.3. Mirror Extension of m-Grids

The subgrid method can be extended to orbifolds when at most two Wigner polynomials with index pairs $[(m_1, m_2), (-m_1, m_2)]$ or linearly combine into an invariant basis polynomial. We call these two cases vertical and horizontal mirror pairs. It then suffices to mark one partner of this pair and to give the algebraic phase for its partner. These cases apply for the orbifolds *N*2 and *N*3, illustrated in Figure 3.

12.4. Projection of Harmonic Bases

For more general deck groups, we establish the identity representation by application to a Wigner polynomial D^j of degree 2*j* the projection operator P^0 Equation (40) to the identity representation of the deck group *H*,

$$P^0 = \frac{1}{|H|} \sum_{g \in H} T_g \tag{40}$$

By Equation (18), this application produces for any degree 2*j* a linear combination of Wigner polynomials. If this can be normalized, it contributes to the harmonic analysis. For Platonic spherical manifolds, these projections were carried out in [22]. In many cases, it is possible to display the results of the projection Equation (40) on an m-subgrid: the non-vanishing Wigner polynomials selected by projection are located on a subgrid, dependent on the manifold, and spanned by two specific grid vectors (a_1, a_2). We display subgrids in Figure 3. The advantage of the subgrid method is that different topologies can be directly compared in terms of their subgrids.

Summary: The Wigner polynomials invariant under a single-generator deck group are located on a subgrid spanned by two grid vectors (a_1, a_2). Any subgrid point (m_1, m_2) stands for an enumerable set of invariant Wigner polynomials of degree $j = j_0 + \nu$, $j_0 = max(|m_1|, |m_2|)$. The spherical manifolds $N1$, $N2$ with cyclic deck groups C_5, C_8 belong to this set.

Table 4. The lowest cubic invariant spherical harmonics YΓ1,lYΓ1,l, expressed by spherical harmonics YlmYml.

l	$Y^{\Gamma_1,l} = \sum_m a_{lm} Y^l_m(\theta, \phi)$
0	Y^0_0
4	$\sqrt{\frac{7}{12}} Y^4_0 + \sqrt{\frac{5}{24}} (Y^4_4 + Y^4_{-4})$
6	$\sqrt{\frac{1}{72}} Y^6_0 - \sqrt{\frac{7}{144}} (Y^6_4 + Y^6_{-4})$
8	$\frac{1}{64}\sqrt{33} Y^8_0 + \frac{1}{12}\sqrt{\frac{21}{2}} (Y^8_4 + Y^8_{-4}) + \frac{1}{24}\sqrt{\frac{195}{2}} (Y^8_8 + Y^8_{-8})$

12.5. Algebraic Bases from Point Symmetry

The basis of the harmonic analysis for a specific topology can be recursively computed, as we shall show. We have seen that the deck group of the orbifold factorizes into the symmetry group M of the Platonic manifold and its deck group H. The basis functions for the harmonic analysis must be invariant under the deck group $M \cdot$

H, which is possible only if they are invariant under both M and H. The two projectors for the groups M and H commute [24].

We start for the cubic orbifold $N3$ the first loop of a recursive construction of basis functions of increasing degree $2j$:

i. The action of symmetry groups $M = O$ of the Platonic manifolds is known from the Euclidean setting in E^3. The point group O is a subgroup of the rotation group $SO(3, R)$; therefore, any M-invariants must appear for fixed l as linear combinations of spherical harmonics ∑malmYlm. Standard spherical harmonics YlmYml with lowest label $l = l_0$ and invariant under O can be found in [25] (pp. 436–438) for the cubic group and in [26] (pp. 105–114). For several point groups M, the onset l_0 of spherical harmonics invariant under M is listed in [10], Table 3. The linear combinations of lowest $l = l_0$ can be expressed in the spherical basis Equation (20) of $SO(4, R)$. For the cubic point group O, the lowest linear combination of spherical harmonics given in Table 4 has $l_0 = 4$.

ii. To also project on an invariant under H, which is $H = Q$ for the orbifold $N3$, we transform with Equation (21) from the spherical basis back to the Wigner basis and then apply the projector Equation (40) with the matrix element:

$$\langle jm_1'm_2'|\frac{1}{|H|}\sum_{(g_l,g_r)\in H} T_{(g_l,g_r)}|jm_1m_2\rangle \tag{41}$$

Once the result of (ii) is transformed back into the spherical basis with Equation (20), we recover linear combinations with new values of multipole order $l > l_0$, which, if non-vanishing, can be normalized. These new states must still be invariant under the point group M. It follows that they must contain M-invariant linear combination of spherical harmonics with $l > l_0$. For the cubic point group O, we demonstrate this result in Table 4 with new cubic invariants for $l = 6$ and $l = 8$.

With them, we enter the next loop with Steps (i) and (ii) in search of new invariant basis functions for the deck group $Q \times_s O$. This recursive method uses only well-known Wigner coefficients of $SU(2, C)$, but avoids any new projection to invariants under M. The new basis functions in Table 5 contain linear combinations of different multipole orders $l = 4, 6, 8$. This implies that topology can enforce linear correlations between amplitudes of different multipole order, which has important implications for the analysis of the cosmic microwave background.

Table 5. The lowest ($Q \times O$)-invariant polynomials ψ0,Γ1,2jψ0,Γ1,2jof degree $2j$ on the three-sphere for the orbifold $N3$, expressed by the cubic invariant spherical harmonics fromTable 4. ($Q \times_s O$)-invariance enforces coherent superpositions of several cubic invariant spherical harmonics.\

$2j$	l	$\psi^{0,\Gamma_1,2j} = \sum_l b_l R_{2j+1\,l}(\chi) Y^{\Gamma_1,l}(\theta,\phi)$
0	0	$R_{10} Y^{\Gamma_1,0}$
4	0, 4	$\sqrt{\frac{2}{5}} R_{50} Y^{\Gamma_1,0} + \sqrt{\frac{3}{5}} R_{54} Y^{\Gamma_1,4}$
6	0, 4, 6	$\sqrt{\frac{1}{7}} R_{70} Y^{\Gamma_1,0} - \sqrt{\frac{6}{11}} R_{74} Y^{\Gamma_1,4} - \sqrt{\frac{24}{77}} R_{76} Y^{\Gamma_1,6}$
8	0, 4, 6, 8	$\frac{4}{3}\sqrt{\frac{1}{110}} R_{90} Y^{\Gamma_1,0} - \frac{12}{11}\sqrt{\frac{3}{65}} R_{94} Y^{\Gamma_1,4}$
		$+\frac{8 \cdot 19}{165} R_{96} Y^{\Gamma_1,6} + \frac{4}{5}\sqrt{\frac{1}{33 \cdot 13}} R_{98} Y^{\Gamma_1,8}$

In Table 5, we exemplify the first basis functions for the cubic orbifold $N3$.

Summary: The recursive method, by use of loops with Steps (i) and (ii), yields a basis of the harmonic analysis of any Platonic orbifold. Invariance under a topological deck group will imply in general the correlation between multipoles of different orders (l, l').

13. SYNOPSIS OF HARMONIC BASES FOR PLATONIC ORBIFOLDS

Depending on the orbifold, we apply one of the methods given in the previous section to derive the basis of each Platonic orbifold.

13.1. The Tetrahedral Orbifold N1

The deck group of *N*1 from [22] is the cyclic group $H = C_5$ of order five. Its single generator is the rotation:

$$T_{(W_1 W_2 W_3 W_4)} = T_{(g_l, g_r)} : (g_l, g_r) = (v_1 v_2^{-1} v_3 v_4^{-1}, v_1^{-1} v_2 v_3^{-1} v_4) \tag{42}$$

where the vector a_i of the Weyl reflection Wi:=WaiWi:=Wai of the Coxeter group from Table 2 by Equation (28) determines the unitary matrix $v_i = v(a_i)$, $i = 1, 2, 3, 4$ (Equation (27)), and Equation (28) is used to convert products of two Weyl reflection operators into rotations.

Equation (42) demonstrates that the entanglement of left and right actions in the group C_5 is enforced by the geometry of the tetrahedron.

We follow Section 3 and replace the old coordinates $u(x)$ by the new ones $u(\tilde{x}) = q^{-1}u(x)k.$ By the trace relationEquation (35) for (g_l, g_r), we find the diagonal forms:

$$\delta_l = \begin{bmatrix} \exp(i\,2\pi/5) & 0 \\ 0 & \exp(-i\,2\pi/5) \end{bmatrix}, \quad \delta_r = \begin{bmatrix} \exp(i\,6\pi/5) & 0 \\ 0 & \exp(-i\,6\pi/5) \end{bmatrix} \tag{43}$$

To project a basis state invariant under C_5, it suffices to make it invariant under the generator Equation (42) of C_5, since then, it will be invariant under any power of this generator. We choose a fixed Wigner polynomial and find in the coordinates u(x˜) as action the phase transformation:

$$(T_{(\delta_l,\delta_r)} D^j_{m_1,m_2})(u(\tilde{x}) = D^j_{m_1,m_2}(\tilde{u}) \exp(i(-m_1 2\pi/5 + m_2\, 6\pi/5) \quad (44)$$

To have invariance under the generator of C_5, we require:

$$-m_1 + 3m_2 \equiv 0 \bmod 5 \quad (45)$$

This condition is fulfilled on the subgrid points marked in Figure 3, *N*1 by black and by red circles. The subgrids are spanned by the black vectors $a_1 = (2, -1)$, $a_2 = (1, 2)$ from an integer or by the same red vectors from a half-integer grid point. The subgrid points form a selection from all grid points. The subbasis consists of all Wigner polynomials associated with the subgrid points.

13.2. The Cubic Manifold N2

The deck group is C_8 generated by:

$$(g_l, g_r) : g_l = \begin{bmatrix} -\overline{a} & 0 \\ 0 & -a \end{bmatrix}, \; g_r = \begin{bmatrix} 0 & \overline{a} \\ -a & 0 \end{bmatrix}, \; a = \exp(i\, 2\pi/8) \quad (46)$$

Again, the left and right actions are entangled.

The subbasis of the harmonic analysis is of the horizontal mirror type (see Section 12.3) given by the linear combination of two Wigner polynomials:

$$\Phi^j_{m_1,0} = D^j_{m_1,0}(u),\ j \in Z,$$
$$\Phi^j_{m_1,m_2} = \frac{1}{\sqrt{2}}[D^j_{m_1,m_2}(u) + i^{(m_1+m_2)}(-1)^{(j+m_2)}D^j_{m_1,-m_2}(u)],$$
$$j \in Z, \text{ for } m_1 \text{ even}, 0 < m_2 \leq j$$

(47)

13.3. The Cubic Manifold N3

The deck group is the quaternionic group Q, acting from the left, with elements:

$$g_l = -\mathbf{k} = \begin{bmatrix} 0 & -i \\ -i & 0 \end{bmatrix},\ g_l = -\mathbf{j} = \begin{bmatrix} 0 & -1 \\ 1 & 0 \end{bmatrix},\ g_l = -\mathbf{i} = \begin{bmatrix} -i & 0 \\ 0 & i \end{bmatrix}$$

(48)

The subbasis is of vertical mirror type, given by a linear combination of two Wigner polynomials:

$$\Phi^j_{m_1,m_2} = \frac{1}{\sqrt{2}}[D^j_{m_1,m_2}(u) - D^j_{-m_1,m_2}(u)] \quad (49)$$

$$\text{for } j \text{ odd}, \geq 3, \text{ for } m_1 \text{ even}, 0 < m_1 \leq j, -j \leq m_2 \leq j$$

(50)

13.4. The Octahedral Manifolds N4, N5, N6

The deck groups here are of the order 48. We give the basis construction to the orbifold $N4$ with deck group Cl3×Qr acting from the left and from the right, respectively. For the subbasis in new coordinates u(x˜), we find a horizontal mirror symmetry and, from [10] (p. 26), the following polynomials and mirror phases:

$$\begin{aligned}
m_1 = \rho \equiv 0 \bmod 3, & \\
\text{for } j \text{ odd}, \geq 3,\ 0 < m_2 \leq j: \quad \Phi^j_{\rho,m_2} &= [D^j_{\rho,m_2}(u(\tilde{x})) - D^j_{\rho,-m_2}(u(\tilde{x}))], \\
\text{for } j \text{ even}, m_2 = 0: \quad \Phi^j_{\rho,0} &= D^j_{\rho,0}(u(\tilde{x})), \\
\text{for } j \text{ even}, \geq 2,\ 0 < m_2 \leq j: \quad \Phi_{\rho,m_2} &= [D^j_{\rho,m_2}(u(\tilde{x})) + D^j_{\rho,-m_2}(u(\tilde{x}))]
\end{aligned} \tag{51}$$

The octahedral manifolds *N*5, *N*6 are analyzed in [10].

13.5. The Dodecahedral Manifold N9

An analysis of invariant polynomials on Poincaré's dodecahedral manifold is given in [2], based on the work of Klein [3].

14. TOPOLOGY OF MULTIPLY-CONNECTED UNIVERSES

As an application of the derived subbases, we consider the observable effects of multiply-connected universes. The topology of the cosmos has its roots in Albert Einstein's work [27]. In 1917, Einstein presented his pioneering paper on general relativity, cosmology and gravitation [1] (pp. 160–164). He communicates three fundamental ideas:

i. Space-time manifold and metric: Space and time are unified into a single 4D space-time manifold, with three space coordinates (x_1, x_2, x_3) and one time coordinate $x_0 = ct$. Following Riemann, this four-manifold carries a pseudo-Euclidean metric with the space-time squared distance:

$$
\begin{aligned}
ds^2 &= \sum_{\mu,\nu=0}^{3} g_{\mu\nu}(x_0, x_1, x_2, x_3)dx_\mu dx_\nu \\
&= (ct)^2 - \sum_{i,j=1}^{3} g_{ij}dx_i dx_j,
\end{aligned} \tag{52}
$$

$$
g_{\mu\nu} = g_{\nu\mu} \tag{53}
$$

In the second line, we adopt the usual splitting into time and space. Space-time by Equation (52) is distinct from E^4used in previous sections.

ii. Gravity and geodesics: Einstein gives field equations for Newton's gravity. They link the second derivatives of the metric tensor $g_{\mu\nu}(x_0, x_1, x_2, x_3)$ with respect to the coordinates x_μ linearly to the energy-momentum tensor $T_{\sigma\rho}(x_0, x_1, x_2, x_3)$. Geodesics, the shortest lines allowed by the metric Equation (52), are the lines followed by massive test particles. For velocities small compared with the velocity c of light, Einstein's field equations reduce to Newton's laws for the gravitational potential in the presence of masses and the differential equations for geodesics to Newton's equations of motion. In general, the metric and the energy-momentum tensor with Einstein become observables of astrophysics.

iii. Topology: The connectivity of the space-time manifold is the subject of cosmic topology. The classes of possible closed paths are the basic concept of homotopy. The view of space-time as a manifold, with gravitation obtained from local differential equations, presents topology as an observable.

When time is split off as in Equation (52), homotopy refers to the $3D$ space part of space-time. In cosmology, three possible curvatures of the universe are distinguished: hyperbolic space with negative, Euclidean space with zero and spherical space with positive average curvature. The present astrophysical data favor positive curvature and, hence, a spherical spatial topology, as discussed here. One important observable is the spatial fluctuations of the cosmic microwave background.

The cosmic microwave background (CMB) is an observed uniform thermal black-body radiation of a present temperature of 2.725 Kelvin with a peak frequency of 160.2 GHz, discovered in 1964 by A. Penzias and R. Wilson [28]. According to big bang cosmology, its origin is dated back to the early epoch of photon decoupling, when neutral atoms were formed at very high density and temperature. Its amplitude today is observed by the Wilkinson probe and up to 2013 by the Planck satellite [29]. After a non-trivial cleaning of the measured temperature for global astrophysical influences, such as the Wolfe–Sachs effect and the local foreground of the solar system and our galaxy, the CMB spatial temperature fluctuations are expanded in multipoles, shown for the lowest values of *l* in Figure 4.

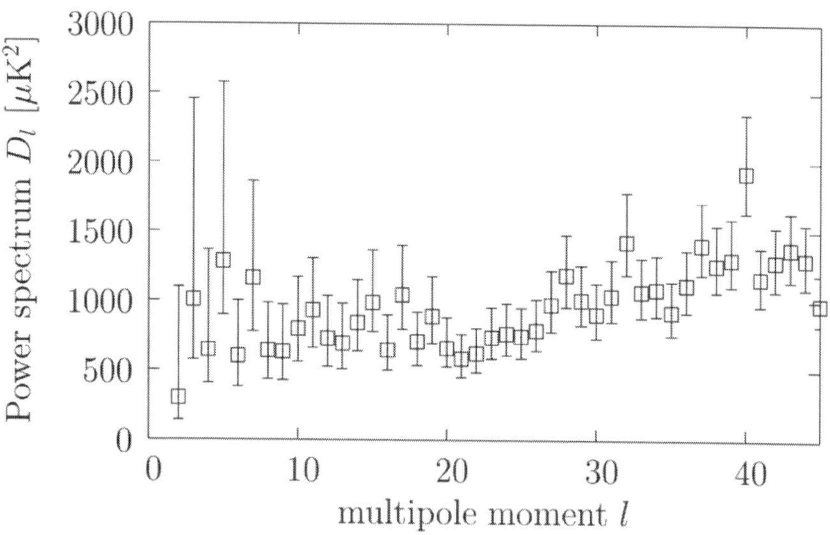

Figure 4. CMB temperature power-spectrum as a function of multipole order *l*; data from [29] (Figure 35). Note the weak amplitudes of some of the lowest multipoles.

The connection of multipole-resolved CMB measurements to cosmic topology is provided by selection rules for the harmonic analysis. The different topologies discussed here should imprint their selection rules on the multipole amplitudes of the CMB. The cubic spherical manifold $N2$ discussed here has been used for detailed predictions of the CMB temperature fluctuation map [30]. In addition, the ratio of different multipole-moments encodes the specific topology; see Table 5 for the cubic case.

Both topological signatures can be revealed with the methods given here, which yields regular tilings of the three-sphere. According to [29], Section 9.3 on p. 36, the observed low amplitudes and correlations of the first CMB multipoles (Figure 4) might indicate a violation of statistical isotropy or might indicate the presence of topological selection rules.

Besides regular tilings, also random tilings of spherical three-manifolds have been analyzed in the literature [31].

15. CONCLUSIONS

We have constructed the orthonormal basis sets for four Platonic spherical topological orbifolds, required for performing a harmonic analysis on specific topologies. The mathematical framework used starts from the homotopy or fundamental groups and yields isomorphic deck actions and deck groups on the three-sphere. The point symmetry groups M of the Platonic manifolds lead from manifolds to orbifolds. The orthonormal Wigner polynomials provide a basis for the harmonic analysis on $SU(2, C)$. By deriving the subgrid of each Platonic manifold on the space spanned by the indices of the Wigner polynomials, we obtain selection rules for the harmonic analysis. Moreover, the coefficients of the multipole expansion encode the underlying topology.

In contrast to other approaches, no numerical orthogonalization of basis sets is required, and the connection between topology and geometry of space is transparent from the introduction of the homotopies. Finally, we discussed the harmonic analysis of the CMB and present signatures of the underlying topology of multiply-connected universes encoded in the CMB fluctuations.

CONFLICTS OF INTEREST

The author declares no conflict of interest.

REFERENCES

1. Einstein, A. Kosmologische Betrachtungen zur allgemeinen Relativitätstheorie. In *Sitzungsberichte der Königlich Preußischen Akademie der Wissenschaften (Berlin)*; Prussian Academy of Sciences: Berlin, Germany, 1917; pp. 142–152.
2. Kramer, P. An invariant operator due to F Klein quantizes H Poincaré's dodecahedral 3-manifold. *J. Phys. A Math. Gen.* **2005**, *38*.
3. Klein, F. *Vorlesungen Über das Ikosaeder*; Birkhäuser: Basel, Switzerland, 1993.
4. Everitt, B. 3-Manifolds from Platonic solids. *Topol. Appl.* **2004**, *138*, 253–263.
5. Seifert, H.; Threlfall, W. *Lehrbuch der Topologie*; Chelsea Reprint: New York, NY, USA, 1934.
6. Wolf, J.A. *Spaces of Constant Curvature*\, 5th ed; Publish or Perish: Wilmington, DE, USA, 1984.
7. Thurston, W.P. *Three-Dimensional Geometry and Topology*; Princeton University Press: Princeton, NJ, USA, 1997.

8. Ratcliffe, J.G. *Foundations of Hyperbolic Manifolds*; Springer: Berlin, Germany, 1994.

9. Montesinos, J.M. *Classical Tesselations and Three-Manifolds*; Springer: Berlin, Germany, 1987.

10. Kramer, P. Platonic topology and CMB fluctuations: Homotopy, anisotropy and multipole selection rules. *Class. Quant. Grav.* **2010**, *27*.

11. Edmonds, A.R. *Angular Momentum in Quantum Mechanics*; Princeton University Press: Princeton, NJ, USA, 1957.

12. Wigner, E.P. *Group Theory and Its Application to the Quantum Mechanics of Atomic Spectra*; Wiley: New York, NY, USA, 1959.

13. Bargmann, V. Group representations on Hilbert spaces of analytic functions. In *Analytic Methods in Mathematical Physics*; Gordon and Breach: New York, NY, USA, 1968.

14. Levin, J. Topology and the cosmic microwave background. *Phys. Rep.* **2002**, *365*, 251–333.

15. Aurich, R.; Janzer, H.S.; Lustig, S.; Steiner, F. Do we live in a 'small universe'? *Class. Quant. Grav.* **2008**, *25*.

16. Magnus, W.; Karrass, A.; Solitar, D. *Combinatorial Group Theory*; Dover: New York, NY, USA, 1966.

17. Humphreys, J.E. *Reflection Groups and Coxeter Groups*; Cambridge University Press: Cambridge, UK, 1990.

18. Sommerville, D.M.Y. *An Introduction to the Geometry of N Dimensions*; Dover: New York, NY, USA, 1958.

19. Coxeter, H.S.M.; Moser, W.O.J. *Generators and Relations for Discrete Groups*; Springer: New York, NY, USA, 1965.

20. Kramer, P. Platonic polyhedra tune the 3-sphere: Harmonic analysis on simplices. *Phys. Scr.* **2009**, *79*.

21. Kramer, P. Platonic polyhedra tune the three-sphere: II. Harmonic analysis on cubic spherical three-manifolds. *Phys. Scr.* **2009**, *80*.

22. Kramer, P. Platonic polyhedra tune the three-sphere: III. Harmonic analysis on octahedral spherical three-manifolds.*Phys. Scr.* **2010**, *81*.

23. Kramer, P. Corrigendum: Platonic polyhedra tune the 3-sphere: Harmonic analysis on simplices. *Phys. Scr.* **2010**, *81*.

24. Kramer, P. Multipole analysis in cosmic topology. In *Symmetries in Nature: Symposium in Memoriam Marcos Moshinsky*; American Institute of Physics (AIP) Publishing: College Park, MD, USA, 2010; Volume 1323, pp. 164–177.

25. Lax, M. *Symmetry Principles in Solid State and Molecular Physics*; Wiley: New York, NY, USA, 1974.

26. Knox, R.S.; Gold, A. *Symmetry in the Solid State*; W.A. Benjamin: New York, NY, USA, 1964.

27. Luminet, J.P. Cosmic topology: Twenty years after. *Gravit. Cosmol.* **2014**, *20*, 18–20.

28. Penzias, A.A.; Wilson, R. Measurement of excess antenna temperature at 4080 Mc/s. *Astrophysical Journal* **1965**, *142*, 419–421.

29. Planck Collaboration. Planck 2013 results. XV. CMB power spectra and likelihood. *Astron. Astrophys.* **2013**, *571*, A15.

30. Aurich, R.; Kramer, P.; Lustig, S. Cosmic microwave background radiation in an inhomogeneous spherical space.*Phys. Scr.* **2011**, *84*.

31. Escudero, J.G. Random tilings of spherical 3-manifolds. *J. Geom. Phys.* **2008**, *58*, 1451–1464.

9

CHAPTER

Harmonic Maps and Biharmonic Maps

Hajime Urakawa

Institute for International Education, Tohoku University, Kawauchi 41, Sendai 980-8576, Japan

ABSTRACT

This is a survey on harmonic maps and biharmonic maps into (1) Riemannian manifolds of non-positive curvature, (2) compact Lie groups or (3) compact symmetric spaces, based mainly on my recent works on these topics.

Keywords: harmonic map; biharmonic map; Chen's conjecture; generalized Chen's conjecture; Lie group; symmetric space

1. INTRODUCTION

Harmonic maps play a central role in geometry; they are critical points of the energy functional $E(\varphi) = \frac{1}{2}\int_M |d\varphi|^2 v_g$ for smooth maps φ of (M, g) into (N, h). The Euler–Lagrange equations are given by the vanishing of the tension field $\tau(\varphi)$. In 1983, J. Eells and L. Lemaire extended [1] the notion of harmonic maps to biharmonic maps, which are, by definition, critical points of the bienergy functional:

$$E_2(\varphi) = \frac{1}{2}\int_M |\tau(\varphi)|^2 v_g \tag{1}$$

After G.Y. Jiang studied [2] the first and second variation formulas of E_2, extensive studies in this area have been done (for instance, see [3–42]). Notice that harmonic maps are always biharmonic by definition.

The outline of this survey is the following:

1. Preliminaries.
2. Chen's conjecture and the generalized Chen's conjecture.
3. Outline of the proofs of Theorems 3–5.
4. Harmonic maps and biharmonic maps into compact Lie groups or compact symmetric spaces.
5. The *CR* analogue of harmonic maps and biharmonic maps.
6. Biharmonic hypersurfaces of compact symmetric spaces.

2. PRELIMINARIES

In this section, we prepare materials for the first and second variational formulas for the bienergy functional and biharmonic maps. Let us recall the definition of a harmonic map φ: $(M, g) \to (N, h)$, of a compact Riemannian manifold (M,g) into another Riemannian manifold (N, h), which is an extremal of the energy functional defined by:

$$E(\varphi) = \int_M e(\varphi)\, v_g$$

where $e(\varphi) := \frac{1}{2}|d\varphi|^2$ is called the energy density of φ. That is, for any variation $\{\varphi_t\}$ of φ with $\varphi_0 = \varphi$,

$$\left.\frac{d}{dt}\right|_{t=0} E(\varphi_t) = -\int_M h(\tau(\varphi), V) v_g = 0 \qquad (2)$$

where $V \in \Gamma(\varphi^{-1}TN)$ is a variation vector field along φ, which is given by $V(x) = \frac{d}{dt}|_{t=0}\varphi_t(x) \in T_{\varphi(x)}N, (x \in M).$, and the tension field is given by $\tau(\varphi) = \sum_{i=1}^m B(\varphi)(e_i, e_i) \in \Gamma(\varphi^{-1}TN).$,

where {ei}mi=1{ei}i=1m is a locally-defined frame field on (M, g), and $B(\varphi)$ is the second fundamental form of φ defined by:

$$\begin{aligned} B(\varphi)(X,Y) &= (\widetilde{\nabla} d\varphi)(X,Y) \\ &= (\widetilde{\nabla}_X d\varphi)(Y) \\ &= \overline{\nabla}_X (d\varphi(Y)) - d\varphi(\nabla_X Y) \end{aligned} \tag{3}$$

for all vector fields X,Y∈X(M). Here, ∇ and ∇^N are connections on *TM*, *TN* of (M, g), (N, h), respectively, and $\overline{\nabla}$ and $\widetilde{\nabla}$ are the induced ones on $\varphi^{-1}TN$ and $T^*M \otimes \varphi^{-1}TN$, respectively. By Equation (2), φ is harmonic if e and only if $\tau(\varphi) = 0$.

The second variation formula is given as follows. Assume that φ is harmonic. Then,

$$\left.\frac{d^2}{dt^2}\right|_{t=0} E(\varphi_t) = \int_M h(J(V), V) v_g \tag{4}$$

where *J* is an elliptic differential operator, called the Jacobi operator acting on $\Gamma(\varphi^{-1}TN)$ given by:

$$J(V) = \overline{\Delta} V - \mathcal{R}(V) \tag{5}$$

where $\overline{\Delta} V = \overline{\nabla}^* \overline{\nabla} V = -\sum_{i=1}^m \{\overline{\nabla}_{e_i} \overline{\nabla}_{e_i} V - \overline{\nabla}_{\nabla_{e_i} e_i} V\}$ is the rough Laplacian and R$\mathcal{R}$ is a linear operator on $\Gamma(\varphi^{-} TN)$ given by $\Gamma(\varphi^{-1}TN)$ given by $\mathcal{R}(V) = \sum_{i=1}^m R^N(V, d\varphi(e_i)) d\varphi(e_i)$,

and R^N is the curvature tensor of $R^N(U,V) = \nabla^N{}_U \nabla^N{}_V - \nabla^N{}_V \nabla^N{}_U - \nabla^N{}_{[U,V]}$ for $U, V \in \mathfrak{X}(N)$.

J. Eells and L. Lemaire [1] proposed polyharmonic (*k*-harmonic) maps, and Jiang [2] studied the first and second variation formulas of biharmonic maps. Let us consider the bienergy functional defined by:

$$E_2(\varphi) = \frac{1}{2}\int_M |\tau(\varphi)|^2 v_g \tag{6}$$

where $|V|^2 = h(V, V), V \in \Gamma(\varphi^{-1}TN)$.

Then, the first variation formula of the bienergy functional is given (the first variation formula) by:

$$\left.\frac{d}{dt}\right|_{t=0} E_2(\varphi_t) = -\int_M h(\tau_2(\varphi), V) v_g \tag{7}$$

Here,

$$\tau_2(\varphi) := J(\tau(\varphi)) = \overline{\Delta}(\tau(\varphi)) - \mathcal{R}(\tau(\varphi)) \tag{8}$$

which is called the bitension field of φ, and J is given in Equation (5).

A smooth map φ of (M, g) into (N, h) is said to be biharmonic if $\tau_2(\varphi) = 0$.

3. CHEN'S CONJECTURE AND THE GENERALIZED CHEN'S CONJECTURE

Recall the famous Chen's conjecture on biharmonic submanifold of the Euclidean space:

Chen's conjecture: A biharmonic submanifold in the Euclidean space must be minimal.

One can consider biharmonic submanifolds of a Riemannian manifold of non-positive curvature, and the generalized Chen's conjecture is the following (*cf.*, R. Caddeo, S. Montaldo, P. Piu [7] and also S. Montaldo, C. Oniciuc [20], *etc.*):

The generalized Chen's conjecture: A biharmonic submanifold in a Riemannian manifold of non-positive curvature must be minimal.

Notice that the generalized Chen's conjecture was solved negatively by giving a counter example by Ou and Tang [27,41]. We first give several comments on Chen's conjecture. It should be emphasized that Chen's conjecture has been still unsolved until now.

Second, we will treat the generalized Chen's conjecture. K. Akutagawa and S. Maeta [3] gave a remarkable breakthrough to Chen's conjecture, by giving the following answer to this conjecture in the case of properly-immersed submanifolds of the Euclidean space:

If we do not assume the properness condition, N. Koiso and myself [43] gave recently a partial answer in the case of generic hypersurfaces of the Euclidean space. Namely, we obtained the following:

Theorem 1. *Let* ι: $(M^n, g) \subset E^{n+1}$ *be an isometrically-immersed biharmonic hypersurface of the Euclidean space. Assume that (1) every principal curvature* λ_i *has multiplicity one, i.e.,* $\lambda_i \neq \lambda_j$ $(i \neq j)$, *and (2) the principal curvature vector* *fields* v_i $(i = 1, \ldots, n)$ *along* ι *satisfy that* $g(\nabla_{v_i} v_j, v_k) \neq 0$ *for all distinct triplets of integers* $i, j, k = 1, \ldots, n$. *Then, it is minimal.*

For harmonic maps, it is well known that:

If a domain manifold (M, g) is complete and has non-negative Ricci curvature and the sectional curvature of a target manifold (N, h) is non-positive, then every energy finite harmonic map is a constant map (*cf.*, [44]).

See [15,45–49] for recent works on harmonic maps. Therefore, it is a natural question to consider biharmonic maps into a Riemannian manifold of non-positive curvature. In this connection, Baird, Fardoun and Ouakkas (*cf.*, [4]) showed that:

If a non-compact Riemannian manifold (M, g) is complete and has non-negative Ricci curvature and (N, h) has non-positive sectional curvature, then every bienergy finite biharmonic map of (M, g) into (N, h) is harmonic.

In our paper [21–23], we showed that:

Theorem 2. *Assume that* (M, g) *is complete and* (N, h) *has non-positive sectional curvature.*

1. *Every biharmonic map* φ: $(M, g) \to (N, h)$ *with finite energy* $E(\varphi) < \infty$ *and finite bienergy* $E_2(\varphi) < \infty$, *is harmonic.*

2. *In the case* $\mathrm{Vol}(M, g) = \infty$, *every biharmonic map* φ: $(M, g) \to (N, h)$ *with finite bienergy* $E_2(\varphi) < \infty$, *is harmonic.*

We do not need any assumption on the Ricci curvature of (M, g) in Theorem 3. If (M, g) is a non-compact complete

Riemannian manifold whose Ricci curvature is non-negative, then $\mathrm{Vol}(M, g) = \infty$ (*cf.*, Theorem 7, p. 667, [50]). Thus, Theorem 3, Equation (2), recovers the result of Baird, Fardoun and Ouakkas. Theorem 3 is sharp, since one cannot weaken the assumptions. Indeed, the generalized Chen's conjecture does not hold if (M, g) is not complete (*cf.*, the counter examples of Ou and Tang [27]). The two assumptions of finiteness of the energy and bienergy are necessary. Indeed, there exists a biharmonic map φ, which is not harmonic, but the energy and bienergy are infinite. For example, $f(x) = r(x)^2 = \sum_{i=1}^{m}(x_i)^2, x = (x_1, \cdots, x_m) \in \mathbb{R}^m$ is biharmonic, but not harmonic, and has infinite energy and bienergy.

As the first bi-product of our method, we obtained (*cf.*, [21,22]):

Theorem 3. *Assume that (M, g) is a complete Riemannian manifold, and let φ: $(M, g) \to (N, h)$ be an isometric immersion; the sectional curvature of (N, h) is non-positive. φ: $(M, g) \to (N, h)$ is biharmonic and $\int_M |\xi|^2 v_g < \infty$, then it is minimal. Here, ξ is the mean curvature normal vector field of the isometric immersion φ.*

Theorem 4 gave an affirmative answer to the generalized B.Y. Chen's conjecture (*cf.*, [7]) under some natural conditions.

For the second bi-product, we can apply Theorem 3 to a horizontally-conformal submersion (*cf.*, [51,52]). Then, we obtain:

Theorem 4. *Let (M^m, g) be a non-compact complete Riemannian manifold $(m > 2)$ and (N^2, h), a Riemannian surface with non-positive curvature. Let λ be a positive* *function on M belonging to $C^\infty(M) \cap L^2(M)$, and φ: $(M, g) \to (N^2, h)$, a horizontally-conformal submersion with a dilation λ. If φ is biharmonic and $\lambda\, |\hat{\mathbf{H}}|_g \in L^2(M)$, then φ is a harmonic morphism. Here, $\hat{\mathbf{H}}$ is trace of the second fundamental form of each fiber of φ.*

4. OUTLINE OF THE PROOFS OF THEOREMS 3–5

In this section, we first give a sketch of the proof of Theorem 3.

The first step: For a fixed point $x_0 \in M$, and for every $0 < r < \infty$, we first take a cut-off C^∞ function η on M satisfying that:

$$\begin{cases} 0 \leq \eta(x) \leq 1 \quad (x \in M), \quad \eta(x) = 1 \quad (x \in B_r(x_0)) \\ \eta(x) = 0 \quad (x \notin B_{2r}(x_0)), \quad |\nabla \eta| \leq \dfrac{2}{r} \quad (x \in M) \end{cases} \tag{9}$$

For a biharmonic map φ: $(M, g) \to (N, h)$, the bitension field is given as:

$$\tau_2(\varphi) = \overline{\Delta}(\tau(\varphi)) - \sum_{i=1}^{m} R^N(\tau(\varphi), d\varphi(e_i))d\varphi(e_i) = 0 \tag{10}$$

so we have:

$$\int_M \langle \overline{\Delta}(\tau(\varphi)), \eta^2 \tau(\varphi)\rangle \, v_g = \int_M \eta^2 \sum_{i=1}^{m} \langle R^N(\tau(\varphi), d\varphi(e_i))d\varphi(e_i), \tau(\varphi)\rangle \, v_g$$
$$\leq 0 \tag{11}$$

since the sectional curvature of (N, h) is non-positive.

The second step: Therefore, by Equation (11) and noticing that $\overline{\Delta}=\overline{\nabla}^*\overline{\nabla}$, we obtain:

$$\begin{aligned}
0 &\geq \int_M \langle \overline{\Delta}(\tau(\varphi)), \eta^2 \tau(\varphi)\rangle \, v_g \\
&= \int_M \langle \overline{\nabla}\tau(\varphi), \overline{\nabla}(\eta^2 \tau(\varphi))\rangle \, v_g \\
&= \int_M \sum_{i=1}^{m} \langle \overline{\nabla}_{e_i}\tau(\varphi), \overline{\nabla}_{e_i}(\eta^2 \tau(\varphi))\rangle \, v_g \\
&= \int_M \sum_{i=1}^{m} \left\{ \eta^2 \langle \overline{\nabla}_{e_i}\tau(\varphi), \overline{\nabla}_{e_i}\tau(\varphi)\rangle + e_i(\eta^2) \langle \overline{\nabla}_{e_i}\tau(\varphi), \tau(\varphi)\rangle \right\} v_g \\
&= \int_M \eta^2 \sum_{i=1}^{m} \left| \overline{\nabla}_{e_i}\tau(\varphi)\right|^2 v_g + 2\int_M \sum_{i=1}^{m} \langle \eta \overline{\nabla}_{e_i}\tau(\varphi), e_i(\eta)\, \tau(\varphi)\rangle \, v_g
\end{aligned} \tag{12}$$

where we used $e_i(\eta^2) = 2\eta\, e_i(\eta)$ at the last equality. By moving the second term in the last equality of Equation (12) to the left-hand side, we have:

$$\int_M \eta^2 \sum_{i=1}^m |\overline{\nabla}_{e_i}\tau(\varphi)|^2 v_g \leq -2\int_M \sum_{i=1}^m \langle \eta\,\overline{\nabla}_{e_i}\tau(\varphi), e_i(\eta)\,\tau(\varphi)\rangle\, v_g$$
$$= -2\int_M \sum_{i=1}^m \langle V_i, W_i\rangle\, v_g$$

(13)

where we put $V_i := \eta\,\overline{\nabla}_{e_i}\tau(\varphi)$ and $W_i := e_i(\eta)\,\tau(\varphi)$ $(i = 1 \cdots, m)$.

Now, recall the following Cauchy–Schwartz inequality:

$$\pm\, 2\,\langle V_i, W_i\rangle \leq \epsilon|V_i|^2 + \frac{1}{\epsilon}|W_i|^2 \qquad (14)$$

for all positive $\epsilon > 0$ because of the inequality $0 \leq |\sqrt{\epsilon}\,V_i \pm \frac{1}{\sqrt{\epsilon}}\,W_i|^2.$. Therefore, for Equation (14), we obtain:

$$-2\int_M \sum_{i=1}^m \langle V_i, W_i\rangle\, v_g \leq \epsilon \int_M \sum_{i=1}^m |V_i|^2\, v_g + \frac{1}{\epsilon}\int_M \sum_{i=1}^m |W_i|^2\, v_g$$

(15)

If we put ∈=12∈=12, we obtain, by Equations (13) and (15),

$$\int_M \eta^2 \sum_{i=1}^m |\overline{\nabla}_{e_i}\tau(\varphi)|^2 v_g \leq \frac{1}{2}\int_M \sum_{i=1}^m \eta^2\,|\overline{\nabla}_{e_i}\tau(\varphi)|^2\, v_g$$
$$+\, 2\int_M \sum_{i=1}^m e_i(\eta)^2\,|\tau(\varphi)|^2\, v_g$$

(16)

Thus, by Equations (12) and (16), we obtain:

$$\int_M \eta^2 \sum_{i=1}^m |\overline{\nabla}_{e_i}\tau(\varphi)|^2 \, v_g \leq 4\int_M |\nabla\eta|^2 \, |\tau(\varphi)|^2 \, v_g \leq \frac{16}{r^2}\int_M |\tau(\varphi)|^2 \, v_g \tag{17}$$

The third step: Since (M, g) is complete and non-compact, we can tend r to infinity. By the assumption $E_2(\varphi) = \frac{1}{2}\int_M |\tau(\varphi)|^2 \, v_g < \infty$, the right-hand side goes to zero. Furthermore, if $r \to \infty$, the left-hand side of Equation (17) goes to $\int_M \sum_{i=1}^m |\overline{\nabla}_{e_i}\tau(\varphi)|^2 \, v_g$ since $\eta = 1$ on $B_r(x_0)$. Thus, we obtain:

$$\int_M \sum_{i=1}^m |\overline{\nabla}_{e_i}\tau(\varphi)|^2 \, v_g = 0 \tag{18}$$

Therefore, we obtain, for every vector field X in M,

$$\overline{\nabla}_X \tau(\varphi) = 0 \tag{19}$$

Then, we have, in particular, $|\tau(\varphi)|$ is constant, say c. Because, for every vector field X on M, at each point in M,

$$X\,|\tau(\varphi)|^2 = 2\langle \overline{\nabla}_X \tau(\varphi), \tau(\varphi)\rangle = 0 \tag{20}$$

Therefore, if $\mathrm{Vol}(M, g) = \infty$ and $c \neq 0$, then:

$$\tau_2(\varphi) = \frac{1}{2}\int_M |\tau(\varphi)|^2 \, v_g = \frac{c^2}{2}\mathrm{Vol}(M, g) = \infty \tag{21}$$

which yields a contradiction. Thus, we have $|\tau(\varphi)| = c = 0$, *i.e.*, φ is harmonic. We have Equation (2).

The fourth step: For Equation (1), assume both $E(\varphi) < \infty$ and $E_2(\varphi) < \infty$. Then, let us consider a one-form α on M defined by:

$$\alpha(X) := \langle d\varphi(X), \tau(\varphi)\rangle, \qquad (X \in \mathfrak{X}(M)) \quad (22)$$

Note here that:

$$\int_M |\alpha|\, v_g = \int_M \left(\sum_{i=1}^m |\alpha(e_i)|^2\right)^{1/2} v_g \leq \int_M |d\varphi|\, |\tau(\varphi)|\, v_g$$
$$\leq \left(\int_M |d\varphi|^2\, v_g\right)^{1/2} \left(\int_M |\tau(\varphi)|^2\, v_g\right)^{1/2} = 2\sqrt{E(\varphi)\, E_2(\varphi)} < \infty \quad (23)$$

Moreover, the divergent $\delta\alpha := -\sum_{i=1}^m (\nabla_{e_i}\alpha)(e_i) \in C^\infty(M)$ turns out (cf., [1], p. 9) as:

$$-\delta\alpha = |\tau(\varphi)|^2 + \langle d\varphi, \overline{\nabla}\tau(\varphi)\rangle = |\tau(\varphi)|^2 \quad (24)$$

Indeed, we have:

$$
\begin{aligned}
-\delta\alpha &= \sum_{i=1}^{m} e_i\langle d\varphi(e_i), \tau(\varphi)\rangle - \sum_{i=1}^{m}\langle d\varphi(\nabla_{e_i}e_i), \tau(\varphi)\rangle \\
&= \langle \sum_{i=1}^{m} \left(\overline{\nabla}_{e_i}(d\varphi(e_i)) - d\varphi(\nabla_{e_i}e_i)\right), \tau(\varphi)\rangle + \sum_{i=1}^{m}\langle d\varphi(e_i), \overline{\nabla}_{e_i}\tau(\varphi)\rangle \\
&= \langle \tau(\varphi), \tau(\varphi)\rangle + \langle d\varphi, \overline{\nabla}\tau(\varphi)\rangle
\end{aligned}
$$

which is equal to $|\tau(\varphi)|$ since $\nabla^{-}\tau(\varphi)=0$

By Equation (24) and $E_2(\varphi) = \frac{1}{2}\int_M |\tau(\varphi)|^2 v_g < \infty$, the function $-\delta\alpha$ is also integrable over M. Thus, together withEquation (23), we can apply Gaffney's theorem (see Theorem 6, below) for the one-form α. By integrating Equation (24) over$M$ and by Gaffney's theorem, we have:

$$0 = \int_M (-\delta\alpha)\, v_g = \int_M |\tau(\varphi)|^2\, v_g \tag{25}$$

which yields that $\tau(\varphi) = 0$. We have Theorem 3. □

Theorem 5. *(Gaffney [53]) Let (M, g) be a complete Riemannian manifold. If a C^1 1-form α satisfies that $\int_M |\alpha|\, v_g < \infty$ and $\int_M (\delta\alpha)\, v_g < \infty$ or,* C^1 vector field X defined by

$(\forall\, Y \in \mathfrak{X}(M))$ satisfies that $\int_M |X|\, v_g < \infty$ and $\int_M \operatorname{div}(X)\, v_g < \infty$, then:

$$\int_M (-\delta\alpha)\, v_g = \int_M \operatorname{div}(X)\, v_g = 0 \tag{26}$$

Our method can be applied to an isometric immersion φ: $(M, g) \rightarrow (N, h)$. In this case, the one-form α defined by Equation (22) in the proof of Theorem 3 vanishes automatically without using Gaffney's theorem, since $\tau(\varphi) = m\ \xi$ belongs to the normal component of $T_{\varphi(x)}N$ $(x \in M)$, where ξ is the mean curvature normal vector field and $m = \dim(M)$. Thus, Equation (24) turns out as:

$$0 = -\delta\alpha = |\tau(\varphi)|^2 + \langle d\varphi, \overline{\nabla}\tau(\varphi)\rangle = |\tau(\varphi)|^2 \quad (27)$$

which implies that $\tau(\varphi) = m\ \xi = 0$, *i.e.*, φ is minimal. Thus, we obtain Theorem 4. □

We also apply Theorem 3 to a horizontally-conformal submersion φ: $(M^m, g) \rightarrow (N^n, h)$ $(m > n \geq 2)$ (*cf.*, [52,54]). In the case that a Riemannian submersion from a space form of constant sectional curvature into a Riemann surface (N^2, h), Wang and Ou (*cf.*, [19,28]) showed that it is biharmonic if and only if it is harmonic. We treat with a submersion from a higher dimensional Riemannian manifold (M, g) (*cf.*, [51]). Namely, let φ: $M \rightarrow N$ be a submersion, and each tangent space T_xM $(x \in M)$ is decomposed into the orthogonal direct sum of the vertical space Vx=Ker(dφx)Vx=Ker(dφx) and the horizontal space H$\mathcal{H}_x$:

$$T_xM = \mathcal{V}_x \oplus \mathcal{H}_x \quad (28)$$

and we assume that there exists a positive C^∞ function λ on M, called the dilation, such that, for each $x \in M$,

$$h(d\varphi_x(X), d\varphi_x(Y)) = \lambda^2(x)\, g(X, Y), \quad (X, Y \in \mathcal{H}_x)$$

(29)

The map φ is said to be horizontally homothetic if the dilation λ is constant along horizontally curves in M.

If $\varphi : (M^m, g) \to (N^n, h)\ (m > n \geq 2)$ is a horizontally-conformal submersion, then, the tension field $\tau(\varphi)$ is given (*cf.*, [51,52]) by:

$$\tau(\varphi) = \frac{n-2}{2}\, \lambda^2\, d\varphi(\mathrm{grad}_{\mathcal{H}}(\frac{1}{\lambda^2})) - (m-n) d\varphi(\hat{\mathbf{H}}) \quad (30)$$

where gradH(1λ2) $\mathrm{grad}_{\mathcal{H}}(\frac{1}{\lambda^2})$ is the H$\mathcal{H}$-component of the decomposition according to Equation (28) of grad(1λ2)grad(1λ2)and $\hat{\mathbf{H}}$ is the trace of the second fundamental form of each fiber, which is given by H^= $\frac{1}{m-n}\sum_{k=n+1}^{m}\mathcal{H}(\nabla_{e_k} e_k)$, where a local orthonormal frame field {ei}mi=1{ei}i=1m on M is taken in such a way that $\{e_{ix}/i = 1, \cdots, n\}$ belong to H$\mathcal{H}_x$ and $\{e_{jx}/j = n+1, \cdots, m\}$ belong to VxVx where x is in a neighborhood in M. Then, due to Theorems 3 and Equation (29), we have immediately:

Theorem 6. *Let (M^m, g) be a complete non-compact Riemannian manifold and (N^n, h) a Riemannian manifold with the non-positive sectional curvature $(m > n \geq 2)$. Let φ: $(M, g) \to (N, h)$ be a horizontally-conformal submersion with the dilation λ satisfying that:*

$$\int_M \lambda^2 \left| \frac{n-2}{2} \lambda^2 \operatorname{grad}_{\mathcal{H}}\left(\frac{1}{\lambda^2}\right) - (m-n)\,\hat{\mathrm{H}} \right|_g^2 v_g < \infty \quad (31)$$

Assume that, either $\int_M \lambda^2 v_g < \infty$ *or* $\mathrm{Vol}(M, g) = \int_M v_g = \infty$. *Then, if* $\varphi\colon (M, g) \to (N, h)$ *is biharmonic, then it is a harmonic morphism.*

Due to Theorem 7, we have:

Corollary 7. *Let* (M^m, g) *be a complete non-compact Riemannian manifold and* (N^2, h) *a Riemannian surface with the non-positive sectional curvature* $(m > n = 2)$. *Let* $\varphi\colon (M, g) \to (N, h)$ *be a horizontally-conformal submersion with the dilation* λ *satisfying that:*

$$\int_M \lambda^2 \left|\hat{\mathrm{H}}\right|_g^2 v_g < \infty \quad (32)$$

Assume that either $\int_M \lambda^2 v_g < \infty$ *or* $\mathrm{Vol}(M, g) = \int_M v_g = \infty$. *Then, if* $\varphi : (M, g) \to (N, h)$ *is biharmonic, then it is a harmonic morphism.*

Corollary 8 implies Theorem 5, immediately. □

Remark 1. *(1) Notice that in Theorem 5, there is no restriction to the dilation* λ *because of* $\dim N = 2$. *This implies that for every positive* C^∞ *function* λ *in* $C^\infty(M) \cap L^2(M)$ *satisfying Equations (31) or (32), we have a harmonic morphism* $\varphi\colon (M^m, g) \to (N^2, h)$.

(2) For a biharmonic map of (M, g) *into* (N, h), *the non-positivity of* (N, h) *implies that:*

$$\langle \tau(\varphi), \overline{\Delta}\tau(\varphi)\rangle = \sum_{i=1}^{m} \langle R^N(\tau(\varphi), d\varphi(e_i))d\varphi(e_i), \tau(\varphi)\rangle \leq 0 \tag{33}$$

which is stronger than the Bochner-type formula $|\tau(\varphi)|\,\Delta|\tau(\varphi)| \geq 0.$ *However, we can prove Theorem 3 in an alternative way by using the latter one. Here,* $\Delta = \sum_{i=1}^{m}(e_i{}^2 - \nabla_{e_i}e_i)$ *denotes the negative Laplace operator acting on* $C^\infty(M)$.

5. HARMONIC MAPS AND BIHARMONIC MAPS INTO COMPACT LIE GROUPS OR SYMMETRIC SPACES

In this section, we treat with harmonic maps and biharmonic maps into compact Lie groups or symmetric spaces of the compact type.

5.1. Biharmonic Maps into Compact Lie Groups

We first treat with harmonic maps and biharmonic maps into compact Lie groups. Let θ be the Maurer–Cartan form on G,*i.e.*, a g-valuedg-valued left invariant one-form on G, which is defined by $\theta_y(Z_y) = Z$, (y∈G,Z∈g)(y∈G,Z∈g). For every C^∞ map ψ of (M, g) into (G, h), let us consider a g-valuedg-valued one-form α on M given by $\alpha = \psi^*\theta$. Then, it is well known (see for example, [55]) that:

Lemma 8. *For every C^∞ map $\psi : (M, g) \to (G, h)$,*

$$\theta(\tau(\psi)) = -\delta\alpha \tag{34}$$

where $\alpha = \psi^*\theta$ and θ is the Maurer–Cartan form of G.

Thus, ψ: $(M, g) \to (G, h)$ is harmonic if and only if $\delta\alpha = 0$.

Furthermore, let $\{X_s\}_{s=1}^n$ be an orthonormal basis of g with respect to the inner product ⟨,⟩. Then, for every $V \in \Gamma(\psi^{-1}TG)$,

$$\begin{aligned} V(x) &= \sum_{i=1}^{n} h_{\psi(x)}(V(x), X_{s\,\psi(x)})\, X_{s\,\psi(x)} \in T_{\psi(x)}G \\ \theta(V)(x) &= \sum_{s=1}^{n} h_{\psi(x)}(V(x), X_{s\,\psi(x)})\, X_s \in \mathfrak{g} \end{aligned} \tag{35}$$

for all $x \in M$. Then, for every X∈X(M),

$$\begin{aligned} \theta(\overline{\nabla}_X V) = \sum_{s=1}^{n} h(\overline{\nabla}_X V, X_s)\, X_s &= \sum_{s=1}^{n} \{X\, h(V, X_s) - h(V, \overline{\nabla}_X X_s)\} X_s \\ &= X(\theta(V)) - \sum_{s=1}^{n} h(V, \overline{\nabla}_X X_s) X_s \end{aligned} \tag{36}$$

where we regarded a vector field $Y \in \mathfrak{X}(G)$ by $Y(x) = Y(\psi(x))\ (x \in M)$ to be an element in the space $\Gamma^{-1}(T\,G)$ of C^∞ sections of $\psi^{-1}TG$.

Here, let us recall that the Levi–Civita connection ∇^h of (G, h) is given (*cf.*, [56,57] Volume II, p. 201, Theorem 3.3) by:

$$\nabla^h_{X_t} X_s = \frac{1}{2}[X_t, X_s] = \frac{1}{2}\sum_{\ell=1}^{n} C_{ts}^{\ell} X_\ell \tag{37}$$

where the structure constant C_{ts}^{ℓ} of ggis defined by $[X_t, X_s] = \sum_{\ell=1}^{n} C_{ts}^{\ell} X_\ell$, and satisfies that:

$$C_{ts}^{\ell} = \langle [X_t, X_s], X_\ell \rangle = -\langle X_s, [X_t, X_\ell] \rangle = -C_{t\ell}^{s} \tag{38}$$

Thus, we have by Equations (37) and (38),

$$\begin{aligned}
\sum_{s=1}^{n} h(V, \overline{\nabla}_X X_s) X_s &= \frac{1}{2}\sum_{s,t=1}^{n} h\left(V, \sum_{\ell=1}^{n} h(\psi_* X, X_t) C_{ts}^{\ell} X_\ell\right) X_s \\
&= -\frac{1}{2}\sum_{s,t,\ell=1}^{n} h(V, X_\ell)\, h(\psi_* X, X_t)\, C_{t\ell}^{s} X_s \\
&= -\frac{1}{2}\sum_{t,\ell=1}^{n} h(V, X_\ell)\, h(\psi_* X, X_t)\, [X_t, X_\ell] \\
&= -\frac{1}{2}\left[\sum_{t=1}^{n} h(\psi_* X, X_t)\, X_t, \sum_{\ell=1}^{n} h(V, X_\ell)\, X_\ell\right] \\
&= -\frac{1}{2}\,[\alpha(X), \theta(V)]
\end{aligned} \tag{39}$$

because we have:

$$\alpha(X) = \theta(\psi_* X) = \sum_{t=1}^{n} h(\psi_* X, X_t) X_t \tag{40}$$

and

$$\theta(V) = \sum_{\ell=1}^{n} h(V, X_\ell)\, \theta(X_\ell) = \sum_{\ell=1}^{n} h(V, X_\ell)\, X_\ell \tag{41}$$

Substituting Equations (40) and (41) into the above, we have Equation (39).

Therefore, we obtain the following together with Equations (36) and (39).

Lemma 9. *For every C^∞ map ψ: $(M, g) \to (G, h)$,*

$$\theta(\overline{\nabla}_X V) = X(\theta(V)) + \frac{1}{2}[\alpha(X), \theta(V)] \qquad (42)$$

where $V \in \Gamma(\psi^{-1}TG)$ and X∈X(M).

Then, we see immediately due to this lemma:

Theorem 10. *For every $\psi \in C^\infty(M, G)$, we have:*

$$\begin{aligned}\theta(\tau_2(\psi)) &= \theta(J(\tau(\psi))) \\ &= -\delta\, d\, \delta\alpha - \mathrm{Trace}_g([\alpha, d\,\delta\alpha])\end{aligned} \qquad (43)$$

where $\alpha = \psi^*\theta$.

Here, let us recall the definition:

Definition 11. *For two*g-valuedg-valued *one-forms α and β on M, we define*g-valuedg-valued *symmetric two-tensor $[\alpha, \beta]$ on M by:*

$$[\alpha, \beta](X, Y) := \frac{1}{2}\{[\alpha(X), \beta(Y)] + [\alpha(Y), \beta(X)]\}, \quad (X, Y \in \mathfrak{X}(M))$$

(44)

and its trace $\mathrm{Trace}_g([\alpha, \beta])$ *by:*

$$\text{Trace}_g([\alpha,\beta]) := \sum_{i=1}^{m}[\alpha,\beta](e_i,e_i) \tag{45}$$

*We also use a*g-valuedg-valued *two-form* $[\alpha \wedge \beta]$ *on M by:*

$$[\alpha \wedge \beta](X,Y) := \frac{1}{2}\{[\alpha(X),\beta(Y)] - [\alpha(Y),\beta(X)]\}, \quad (X,Y \in \mathfrak{X}(M)) \tag{46}$$

Then, we have immediately by Theorem 11:

Corollary 12. *For every* $\psi \in C^\infty(M, G)$, *we have:*

1. $\psi: (M, g) \to (G, h)$ *is harmonic if and only if:*

$$\delta\alpha = 0 \tag{47}$$

2. $\psi: (M, g) \to (G, h)$ *is biharmonic if and only if:*

$$\delta\, d\, \delta\alpha + \text{Trace}_g([\alpha, d\,\delta\alpha]) = 0 \tag{48}$$

5.2. Biharmonic Maps into Compact Symmetric Spaces

Now, let θ be the Maurer–Cartan form on G, *i.e.*, a g-valuedg-valued left invariant one-form on G, which is defined by θy(Zy)=Z(y∈G,Z∈g). For every C^∞ map φ of (M, g) into $(G/K, h)$ with a lift $\psi: M \to G$, let us consider a g-valuedg-valuedone-form α on M given by $\alpha = \psi^*\theta$ and the decomposition:

$$\alpha = \alpha_{\mathfrak{k}} + \alpha_{\mathfrak{m}} \tag{49}$$

corresponding to the decomposition g=t⊕m. Then, it is well known (see, for example, [55]) that:

Lemma 13. *For every C^∞ map φ: $(M, g) \to (G/K, h)$,*

$$t_{\psi(x)^{-1}*}\tau(\varphi) = -\delta(\alpha_{\mathfrak{m}} + \sum_{i=1}^{m}[\alpha_{\mathfrak{k}}(e_i), \alpha_{\mathfrak{m}}(e_i)], \quad (x \in M) \tag{50}$$

where $\alpha = \varphi^*\theta$, θ is the Maurer–Cartan form of G and $\delta(\alpha_{\mathrm{m}})$ is the co-differentiation of the m-*valued one-form α_{m} on (M, g).*

Thus, φ: $(M, g) \to (G/K, h)$ is harmonic if and only if:

$$-\delta(\alpha_{\mathfrak{m}}) + \sum_{i=1}^{m}[\alpha_{\mathfrak{k}}(e_i), \alpha_{\mathfrak{m}}(e_i)] = 0 \tag{51}$$

Furthermore, we obtain:

Theorem 14. *We have:*

$$t_{\psi(x)^{-1}*}\tau_2(\varphi) = \Delta_g\left(-\delta(\alpha_{\mathfrak{m}}) + \sum_{i=1}^{m}[\alpha_{\mathfrak{k}}(e_i), \alpha_{\mathfrak{m}}(e_i)]\right) + \sum_{s=1}^{m}\left[\left[-\delta(\alpha_{\mathfrak{m}}) + \sum_{i=1}^{m}[\alpha_{\mathfrak{k}}(e_i), \alpha_{\mathfrak{m}}(e_i)], \alpha_{\mathfrak{m}}(e_s)\right], \alpha_{\mathfrak{m}}(e_s)\right] \tag{52}$$

where Δ_g is the (positive) Laplacian of (M, g) acting on C^∞ functions on M, and{ei}mi=1{ei}i=1m *is a local orthonormal frame field on (M, g).*

Therefore, we obtain immediately the following two corollaries.

Corollary 15. *Let $(G/K, h)$ be a Riemannian symmetric space and $\varphi: (M, g) \to (G/K, h)$ a C^∞ mapping. Then, we have:*

1. *The map $\varphi: (M, g) \to (G/K, h)$ is harmonic if and only if:*

$$- \delta(\alpha_{\mathfrak{m}}) + \sum_{i=1}^{m} [\alpha_{\mathfrak{k}}(e_i), \alpha_{\mathfrak{m}}(e_i)] = 0 \tag{53}$$

2. *The map $\varphi: (M, g) \to (G/K, h)$ is biharmonic if and only if:*

$$\Delta_g \left(-\delta(\alpha_{\mathfrak{m}}) + \sum_{i=1}^{m} [\alpha_{\mathfrak{k}}(e_i), \alpha_{\mathfrak{m}}(e_i)] \right) + \sum_{s=1}^{m} \left[\left[-\delta(\alpha_{\mathfrak{m}}) + \sum_{i=1}^{m} [\alpha_{\mathfrak{k}}(e_i), \alpha_{\mathfrak{m}}(e_i)] , \alpha_{\mathfrak{m}}(e_s) \right] , \alpha_{\mathfrak{m}}(e_s) \right] = 0 \tag{54}$$

Corollary 16. *Let $(G/K, h)$ be a Riemannian symmetric space and $\varphi: (M, g) \to (G/K, h)$ a C^∞ mapping with a horizontal lift $\psi: M \to G$, i.e., $\varphi = \pi \circ \psi$ and $\psi_x(T_xM) \subset H_{\psi(x)}$, which is equivalent to*αt≡0αt≡0.

Then, we have:

1. *The map $\varphi: (M, g) \to (G/K, h)$ is harmonic if and only if:*

$$\delta(\alpha_{\mathfrak{m}}) = 0 \tag{55}$$

2. *and the map* φ: $(M, g) \to (G/K, h)$ *is biharmonic if and only if:*

$$\delta\, d\, \delta(\alpha_{\mathrm{m}}) + \sum_{s=1}^{m} \left[\left[\delta(\alpha_{\mathrm{m}}), \alpha_{\mathrm{m}}(e_s) \right], \alpha_{\mathrm{m}}(e_s) \right] = 0. \tag{56}$$

For applications and examples, see [29,30].

6. THE *CR* ANALOGUE OF HARMONIC MAPS AND BIHARMONIC MAPS

In the 1970s, Chern and Moser initiated [58] the geometry and analysis of strictly convex *CR* manifolds, and many mathematicians have worked on *CR* manifolds (*cf.*, [59]). Recently, Barletta, Dragomir and Urakawa gave [60] the notion of the pseudo-harmonic map, and also Dragomir and Montaldo settled [10] the one of the pseudo-biharmonic map.

6.1. Conjecture and Results

In this part, we raise:

The *CR* analogue of the generalized Chen's conjecture: Let (M, g_θ) be a complete strictly pseudoconvex *CR* manifold and assume that (N, h) is a Riemannian manifold of non-positive curvature.

Then, every pseudo-biharmonic isometric immersion φ: $(M, g_\theta) \to (N, h)$ must be pseudo-harmonic.

We will see that this conjecture holds under some L^2 condition on a complete strongly pseudoconvex *CR* manifold (*cf.*, Theorem 18) and will give characterization theorems on pseudo-biharmonic immersions from *CR* manifolds into the unit sphere or the complex projective space (*cf.*, Theorems 19 and 20). More precisely, we will see:

Theorem 17. *(cf., Theorem 21) Let φ be a pseudo-biharmonic map of a complete CR manifold (M, g_θ) into a Riemannian manifold(N, h) of non-positive curvature. If the pseudo-energy and the pseudo-bienergy of φ are finite, then φ is pseudo-harmonic.*

Then, we have:

Theorem 18. *Let φ be an isometric immersion of a CR manifold (M^{2n+1}, g_θ) into the unit sphere $S^{2n+2}(1)$ of curvature one. Assume that the pseudo-mean curvature is parallel, but not pseudo-harmonic.*

Then, φ is pseudo-biharmonic if and only if the restriction of the second fundamental form B_φ to the holomorphic subspace $H_x(M)$ of T_xM $(x \in M)$ satisfies that:

$$\| B_\varphi|_{H(M)\times H(M)} \|^2 = 2n$$

Furthermore, we have:

Theorem 19. *Let φ be an isometric immersion of a CR manifold (M^{2n+1}, g_θ) into the complex projective space $(\mathbb{P}^{n+1}(c), h, J)$ of holomorphic sectional curvature $c > 0$. Assume that the pseudo-mean curvature is parallel, but not* *pseudo-harmonic. Then, φ is pseudo-biharmonic if and only if one of the following holds:*

1. *$J(d\varphi(T))$ is tangent to $\varphi(M)$ and:*

$$\| B_{\varphi}|_{H(M)\times H(M)} \|^2 = \frac{c}{4}(2n+3)$$

2. *$J(d\varphi(T))$ is normal to $\varphi(M)$ and:*

$$\| B_{\varphi}|_{H(M)\times H(M)} \|^2 = \frac{c}{4}(2n) = \frac{n}{2}c$$

Here, T is the characteristic vector field of (M, g_θ), $H_x(M) \oplus \mathbb{R}T_x = T_x(M)$ and $B_\varphi/_{H(M)\times H(M)}$ is the restriction of the second fundamental form B_φ to $H_x(M)$ $(x \in M)$.

Several examples of pseudo-biharmonic immersions of (M, g_θ) into the unit sphere or complex projective space are given in [31].

6.2. Explanations of Notions and Proofs of the CR Rigidity

We explain the terminologies in the above results following Dragomir and Montaldo [10] and also Barletta, Dragomir and Urakawa [60]. We also prepare the materials on pseudo-harmonic maps and pseudo-biharmonic maps (see also [61]).

Let M be a strictly pseudoconvex CR manifold of $(2n + 1)$-dimension, T the characteristic vector field on M, J the complex structure of the subspace $H_x(M)$ of $T_x(M)$ $(x \in M)$ and g_θ the Webster–Riemannian metric on M defined for $X, Y \in H(M)$ by:

$$g_\theta(X,Y) = (d\theta)(X,JY), \; g_\theta(X,T) = 0, \; g_\theta(T,T) = 1$$

Let us recall for a C^∞ map φ of (M, g_θ) into another Riemannian manifold (N, h); the pseudo-energy $E_b(\varphi)$ is defined [60] by:

$$E_b(\varphi) = \frac{1}{2}\int_M \sum_{i=1}^{2n} (\varphi^* h)(X_i, X_i)\,\theta \wedge (d\theta)^n \tag{57}$$

where {Xi}2ni=1{Xi}i=12n is an orthonormal frame field on $(H(M), g_\theta)$. Then, the first variational formula of $E_b(\varphi)$ is as follows [60]. For every variation $\{\varphi_t\}$ of φ with $\varphi_0 = \varphi$,

$$\left.\frac{d}{dt}\right|_{t=0} E_b(\varphi_t) = -\int_M h(\tau_b(\varphi), V)\, d\theta \wedge (d\theta)^n = 0 \tag{58}$$

where $V \in \Gamma(\varphi^{-1}TN)$ is defined by $V(x) = \frac{d}{dt}|_{t=0}\varphi_t(x) \in T_{\varphi(x)}N, \ (x \in M)$. Here, $\tau_b(\varphi)$ is the pseudo-tension field, which is given by:

$$\tau_b(\varphi) = \sum_{i=1}^{2n} B_\varphi(X_i, X_i) \tag{59}$$

where Bφ(X,Y)(X,Y∈X(M))Bφ(X,Y)(X,Y∈X(M)) is the second fundamental form of Equation (3) for a C^∞ map of (M, g_θ) into (N, h). Then, φ is pseudo-harmonic if $\tau_b(\varphi) = 0$.

The second variational formula of E_b is given as follows ([60], p.733):

$$\left.\frac{d^2}{dt^2}\right|_{t=0} E_b(\varphi_t) = \int_M h(J_b(V), V)\, \theta \wedge (d\theta)^n \qquad (60)$$

where J_b is a subelliptic operator acting on $\Gamma(\varphi^{-1}TN)$ given by:

$$J_b(V) = \Delta_b V - \mathcal{R}_b(V) \qquad (61)$$

Here, for $V \in \Gamma(\varphi^{-1}TN)$),

$$\begin{cases} \Delta_b V = (\overline{\nabla}^H)^* \overline{\nabla}^H V = -\sum_{i=1}^{2n} \left\{ \overline{\nabla}_{X_i}(\overline{\nabla}_{X_i} V) - \overline{\nabla}_{\nabla_{X_i} X_i} V \right\} \\ \mathcal{R}_b(V) = \sum_{i=1}^{2n} R^h(V, d\varphi(X_i)) d\varphi(X_i) \end{cases} \qquad (62)$$

where ∇ is the Tanaka–Webster connection, $\overline{\nabla}$ the induced connection on $\phi^{-1}TN$ induced from the Levi–Civita connection ∇^h and $\{X_i\}_{i=1}^{2n}$ a local orthonormal frame field on $(H(M), g_\theta)$, respectively. Here, $(\overline{\nabla}^H)_X V := \overline{\nabla}_{X^H} V$ $(X \in \mathfrak{X}(M),\ V \in \Gamma(\phi^{-1}TN))$, corresponding to the decomposition $X = X^H + g_\theta(X, T)T$ $(X^H \in H(M))$. Define $\pi_H(X) = X^H$ $(X \in T_x(M))$,

and (∇¯¯¯H)∗(∇¯H)* is the formal adjoint of $(\overline{\nabla}^H)^*$

Dragomir and Montaldo [10] introduced the pseudo-bienergy given by:

$$E_{b,2}(\varphi) = \frac{1}{2}\int_M h(\tau_b(\varphi), \tau_b(\varphi))\, \theta \wedge (d\theta)^n \tag{63}$$

where $\tau_b(\varphi)$ is the pseudo-tension field of φ. They gave the first variational formula of $E_{b,2}$ as follows ([10], p. 227):

$$\left.\frac{d}{dt}\right|_{t=0} E_{b,2}(\varphi_t) = -\int_M h(\tau_{b,2}(\varphi), V)\, \theta \wedge (d\theta)^n \tag{64}$$

where $\tau_{b,2}(\varphi)$ is called the pseudo-bitension field given by:

$$\tau_{b,2}(\varphi) = \Delta_b(\tau_b(\varphi)) - \sum_{i=1}^{2n} R^h(\tau_b(\varphi), d\varphi(X_i))\, d\varphi(X_i) \tag{65}$$

Then, a smooth map φ of (M, g_θ) into (N, h) is said to be pseudo-biharmonic if $\tau_{b,2}(\varphi) = 0$. By definition, a pseudo-harmonic map is always pseudo-biharmonic.

Theorem 20. *(cf., Theorem 18) Assume that φ is a pseudo-biharmonic map of a strictly pseudoconvex complete CR manifold (M, g_θ)into another Riemannian manifold (N, h) of non-positive curvature.*

If φ has finite pseudo-bienergy $E_{b,2}(\varphi) < \infty$ and finite pseudo-energy $E_b(\varphi) < \infty$, then it is pseudo-harmonic, i.e., $\tau_b(\varphi) = 0$.

(Proof of Theorem 21) The proof is divided into several steps.

The first step: For an arbitrarily fixed point $x_0 \in M$, let $B_r(x_0) = \{x \in M: r(x) < r\}$ where $r(x)$ is a distance function on (M, g_θ), and let us take a cut-off function η on (M, g_θ), *i.e.*,

$$\begin{cases} 0 \leq \eta(x) \leq 1 \quad (x \in M), \quad \eta(x) = 1 \quad (x \in B_r(x_0)) \\ \eta(x) = 0 \quad (x \notin B_{2r}(x_0)), \quad |\nabla^{g_\theta} \eta| \leq \dfrac{2}{r} \quad (x \in M) \end{cases} \tag{66}$$

where r is the distance function and ∇g^θ is the Levi–Civita connection of (M, g_θ), respectively. Assume that φ: $(M, g_\theta) \to (N,h)$ is a pseudo-biharmonic map, *i.e.*,

$$\tau_{b,2}(\varphi) = J_b(\tau_b(\varphi)) = \Delta_b(\tau_b(\varphi)) - \sum_{j=1}^{2n} R^h(\tau_b(\varphi), d\varphi(X_j))\, d\varphi(X_j) = 0 \tag{67}$$

The second step: Then, we have:

$$\int_M \langle \Delta_b(\tau_b(\varphi)), \eta^2\, \tau_b(\varphi)\rangle\, \theta \wedge (d\theta)^n$$
$$= \int_M \eta^2 \sum_{j=1}^{2n} \langle R^h(\tau_b(\varphi), d\varphi(X_j))\, d\varphi(X_j),\ \tau_b(\varphi)\rangle\, \theta \wedge (d\theta)^n \leq 0 \tag{68}$$

since (N, h) has the non-positive sectional curvature. However, for the left-hand side of Equation (68), it holds that:

$$\begin{aligned} &\int_M \langle \Delta_b(\tau_b(\varphi)), \eta^2\, \tau_b(\varphi)\rangle\, \theta \wedge (d\theta)^n \\ &\quad = \int_M \langle \overline{\nabla}^H \tau_b(\varphi), \overline{\nabla}^H (\eta^2\, \tau_b(\varphi))\rangle\, \theta \wedge (d\theta)^n \\ &\quad = \int_M \sum_{j=1}^{2n} \langle \overline{\nabla}_{X_j} \tau_b(\varphi), \overline{\nabla}_{X_j} (\eta^2\, \tau_b(\varphi))\rangle\, \theta \wedge (d\theta)^n \end{aligned} \tag{69}$$

Here, let us recall, for $V, W \in \Gamma(\varphi^{-1}TN))$,

$$\langle \overline{\nabla}^H V, \overline{\nabla}^H W \rangle = \sum_{\alpha=1}^{2n+1} \langle \overline{\nabla}^H_{e_\alpha} V, \overline{\nabla}^H_{e_\alpha} W \rangle = \sum_{j=1}^{2n} \langle \overline{\nabla}_{X_i} V, \overline{\nabla}_{X_i} W \rangle$$

where $\{e_\alpha\}_{\alpha=1}^{2n+1}$ is a locally-defined orthonormal frame field of (M, g_θ), $X_{j\,j=1}^{\;2n}$ is an orthonormal frame of $H(M)$ and $\overline{\nabla}^H_X W \; (X \in \mathfrak{X}(M),\, W \in \Gamma(\varphi^{-1}TN))$

$$\overline{\nabla}^H_X W = \sum_j \{(X^H f_j)\, V_j + f_j \overline{\nabla}_{X^H} V_j\}$$

for $W = \sum_j f_i V_j \; (f_j \in C^\infty(M)$ and $V_j \in \Gamma(\varphi^{-1}TN)$. Here, X^H is the $H(M)$-component of X corresponding to the decomposition of $T_x(M) = H_x(M) \oplus \mathbb{R}T_x \; (x \in M)$, and $\overline{\nabla}$ is the induced connection of $\varphi^{-1}TN$ from the Levi–Civita connection ∇^h of (N, h).

Since

$$\overline{\nabla}_{X_j}(\eta^2\, \tau_b(\varphi)) = 2\eta\, X_j \eta\, \tau_b(\varphi) + \eta^2\, \overline{\nabla}_{X_j} \tau_b(\varphi) \tag{70}$$

the right-hand side of Equation (69) is equal to:

$$\int_M \eta^2 \sum_{j=1}^{2n} |\overline{\nabla}_{X_j}\tau_b(\varphi)|^2 \theta \wedge (d\theta)^n + 2\int_M \sum_{j=1}^{2n} \langle \eta\,\overline{\nabla}_{X_j}\tau_b(\varphi),\ (X_j\eta)\,\tau_b(\varphi)\rangle\,\theta \wedge (d\theta)^n \tag{71}$$

Therefore, together with Equation (68), we have:

$$\begin{aligned}\int_M \eta^2 \sum_{j=1}^{2n} |\overline{\nabla}_{X_j}\,\tau_b(\varphi)|^2 \theta \wedge (d\theta)^n &\leq -2\int_M \sum_{j=1}^{2n} \langle \eta\,\overline{\nabla}_{X_j}\tau_b(\varphi),\ (X_j\eta)\,\tau_b(\varphi)\rangle\,\theta \wedge (d\theta)^n \\ &=: -2\int_M \sum_{j=1}^{2n} \langle V_j, W_j\rangle\,\theta \wedge (d\theta)^n\end{aligned} \tag{72}$$

where we define $V_j, W_j \in \Gamma(\varphi^{-1}TN)\ (j = 1, \cdots, 2n)$ by:

$$V_j := \eta\,\overline{\nabla}_{X_j}\,\tau_b(\varphi), \quad W_j := (X_j\eta)\,\tau_b(\varphi)$$

Then, since it holds that $0 \leq \left|\sqrt{\epsilon}\,V_i \pm \frac{1}{\sqrt{\epsilon}}\,W_i\right|^2$ for every $\epsilon > 0$, , we have,

$$\text{RHS of Equation (72)} \leq \epsilon\int_M \sum_{j=1}^{2n} |V_j|^2 \theta \wedge (d\theta)^n + \frac{1}{\epsilon}\int_M \sum_{j=1}^{2n} |W_j|^2 \theta \wedge (d\theta)^n \tag{73}$$

for every ϵ > 0. By taking = 1, we obtain: 2

$$\int_M \eta^2 \sum_{j=1}^{2n} \left| \overline{\nabla}_{X_j} \tau_b(\varphi) \right|^2 \theta \wedge (d\theta)^n$$
$$\leq \frac{1}{2} \int_M \sum_{j=1}^{2n} \eta^2 \left| \overline{\nabla}_{X_j} \tau_b(\varphi) \right|^2 \theta \wedge (d\theta)^n + 2 \int_M \sum_{j=1}^{2n} \left| X_j \eta \right|^2 \left| \tau_b(\varphi) \right|^2 \theta \wedge (d\theta)^n \quad (74)$$

Therefore, we obtain, due to the properties that η = 1 on $B_r(x_0)$ and $\eta = 1$ on $B_r(x_0)$ and $\sum_{j=1}^{2n} |X_j \eta|^2 \leq |\nabla^{g_\theta} \eta|^2 \leq \left(\frac{2}{r}\right)^2$,

$$\int_{B_r(x_0)} \sum_{j=1}^{2n} \left| \overline{\nabla} \tau_b(\varphi) \right|^2 \theta \wedge (d\theta)^n \leq \int_M \eta^2 \sum_{j=1}^{2n} \left| \overline{\nabla}_{X_j} \tau_b(\varphi) \right|^2 \theta \wedge (d\theta)^n$$
$$\leq 4 \int_M \sum_{j=1}^{2n} \left| X_j \eta \right|^2 \left| \tau_b(\varphi) \right|^2 \theta \wedge (d\theta)^n \leq \frac{16}{r^2} \int_M \left| \tau_b(\varphi) \right|^2 \theta \wedge (d\theta)^n \quad (75)$$

The third step: If we let $r \to \infty$, then $B_r(x_0)$ goes to M, and the right-hand side of Equation (75) goes to zero due to our assumptions that $E_{b,2}(\varphi) = \frac{1}{2} \int_M |\tau_b(\varphi)|^2 \theta \wedge (d\theta)^n < \infty$ and (M, g_θ) is complete. Thus, we have:

$$\int_M \sum_{j=1}^{2n} \left| \overline{\nabla}_{X_j} \tau_b(\varphi) \right|^2 \theta \wedge (d\theta)^n = 0 \quad (76)$$

This implies that:

$$\overline{\nabla}_X \tau_b(\varphi) = 0 \qquad \text{(for all } X \in H(M)) \quad (77)$$

The fourth step: Let us take a one form α on M defined by:

$$\alpha(X) = \begin{cases} \langle d\varphi(X), \tau_b(\varphi)\rangle, & (X \in H(M)) \\ 0 & (X = T) \end{cases}$$

Then, we have:

$$\begin{aligned} \int_M |\alpha|\, \theta \wedge (d\theta)^n &= \int_M \left(\sum_{j=1}^{2n} \alpha(X_j)|^2\right)^{\frac{1}{2}} \theta \wedge (d\theta)^n \\ &\leq \left(\int_M |d_b\varphi|^2\, \theta \wedge (d\theta)^n\right)^{\frac{1}{2}} \left(\int_M |\tau_b(\varphi)|^2\, \theta \wedge (d\theta)^n\right)^{\frac{1}{2}} \\ &= 2\sqrt{E_b(\varphi)\, E_{b,2}(\varphi)} < \infty \end{aligned} \tag{78}$$

where we put $d_b\varphi := \sum_{i=1}^{2n} d\varphi(X_i) \otimes X_i,$

$$|d_b\varphi|^2 = \sum_{i,j=1}^{2n} g_\theta(X_i, X_j)\, h(d\varphi(X_i), d\varphi(X_j)) = \sum_{i=1}^{2n} h(d\varphi(X_i), d\varphi(X_i))$$

and

$$E_b(\varphi) = \frac{1}{2}\int_M |d_b\varphi|^2\ \theta \wedge (d\theta)^n \tag{79}$$

Furthermore, let us define a C^∞ function $\delta_b\alpha$ on M by:

$$\delta_b\alpha = -\sum_{j=1}^{2n} (\nabla_{X_j}\alpha)(X_j) = -\sum_{j=1}^{2n} \{X_j(\alpha(X_j)) - \alpha(\nabla_{X_j}X_j)\} \tag{80}$$

where ∇ is the Tanaka–Webster connection. Notice that:

$$\begin{aligned}\operatorname{div}(\alpha) &= \sum_{j=1}^{2n}(\nabla^{g_\theta}_{X_j}\alpha)(X_j) + (\nabla^{g_\theta}_T\alpha)(T)\\ &= \sum_{j=1}^{2n}\{X_j(\alpha\circ\pi_H(X_j)) - \alpha\circ\pi_H(\nabla^{g_\theta}_{X_j}X_j)\}\\ &\quad + T(\alpha\circ\pi_H(T)) - \alpha\circ\pi_H(\nabla^{g_\theta}_T T)\\ &= \sum_{j=1}^{2n}\{X_j(\alpha(X_j)) - \alpha(\pi_H(\nabla^{g_\theta}_{X_j}X_j))\} = \sum_{j=1}^{2n}\{X_j(\alpha(X_j)) - \alpha(\nabla_{X_j}X_j)\}\\ &= -\delta_b\alpha\end{aligned} \tag{81}$$

where π_H: $T_x(M) \to H_x(M)$ is the natural projection. We used the facts that ∇gθTT=0 and $\pi_H(\nabla^{g_\theta}_X Y) = \nabla_X Y \ (X, Y \in H(M))$ ([61], p. 37). Here, recall again that $\nabla^{g_\theta}_T T = 0$ is the Levi-Civita connection of g_θ and ∇ is the Tanaka-Webster connection. Then, we have, for Equation (80),

$$\begin{aligned}\delta_b\alpha &= -\sum_{j=1}^{2n}\{X_j\langle d\varphi(X_j), \tau_b(\varphi)\rangle - \langle d\varphi(\nabla_{X_j}X_j), \tau_b(\varphi)\rangle\}\\ &= -\sum_{j=1}^{2n}\left\{\begin{array}{l}\langle\overline{\nabla}_{X_j}(d\varphi(X_j)), \tau_b(\varphi)\rangle + \langle d\varphi(X_j), \overline{\nabla}_{X_j}\tau_b(\varphi)\rangle\\ -\langle d\varphi(\nabla_{X_j}X_j), \tau_b(\varphi)\rangle\end{array}\right\}\\ &= -\left\langle\sum_{j=1}^{2n}\{\overline{\nabla}_{X_j}(d\varphi(X_j)) - d\varphi(\nabla_{X_j}X_j)\}, \tau_b(\varphi)\right\rangle = -|\tau_b(\varphi)|^2\end{aligned} \tag{82}$$

We used Equation (77) to derive the last second equality of Equation (82). Then, due to Equation (82), we have for $E_{b,2}(\varphi)$,

$$
\begin{aligned}
E_{b,2}(\varphi) &= \frac{1}{2}\int_M |\tau_b(\varphi)|^2\, \theta\wedge(d\theta)^n = -\frac{1}{2}\int_M \delta_b\alpha\ \theta\wedge(d\theta)^n \\
&= \frac{1}{2}\int_M \operatorname{div}(\alpha)\,\theta\wedge(d\theta)^n = 0
\end{aligned} \tag{83}
$$

In the last equality, we used Gaffney's theorem (*cf.*, Theorem 6, [23], p. 271, or [53]).

Therefore, we obtain $\tau_b(\varphi) \equiv 0$, *i.e.*, φ is pseudo-harmonic. □

7. BIHARMONIC HYPERSURFACES OF COMPACT SYMMETRIC SPACES

7.1. Characterization of Biharmonic Maps

In the first part of this section, we show a characterization theorem for an isometric immersion φ of an m-dimensional Riemannian manifold (M, g) into a Riemannian manifold (N, h) whose tension field $\tau(\varphi)$ satisfies that $\overline{\nabla}^{\perp}_X \tau(\varphi) = 0\ (X \in \mathfrak{X}(M))$ is biharmonic. Let us recall the following theorem due to [2]:

Theorem 21. *Let* φ: $(M^m, g) \rightarrow (N^n, h)$ *be an isometric immersion. Assume that* $\overline{\nabla}^{\perp}_X \tau(\varphi) = 0$ *for all* (X∈X(M)). *Then,* φ *is biharmonic if and only if the following holds:*

$$- \sum_{j,k=1}^{m} h(\tau(\varphi), R^h(d\varphi(e_j), d\varphi(e_k))d\varphi(e_k))\, d\varphi(e_j)$$
$$+ \sum_{j,k=1}^{m} h(\tau(\varphi), B_\varphi(e_j, e_k))\, B_\varphi(e_j, e_k) - \sum_{j=1}^{m} R^h(\tau(\varphi), d\varphi(e_j))\, d\varphi(e_j) = 0$$

(84)

where R^h *is the curvature tensor field of* (N, h) *given by* $R^h(U,V)W = \nabla^h_U(\nabla^h_V W) - \nabla^h_V(\nabla^h_U W) - \nabla^h_{[U,V]}W,\ (U,V,W \in \mathfrak{X}(N)),\ B_\varphi(X,Y)\ (X,Y \in \mathfrak{X}(M))$ *is the second fundamental form of the immersion* φ *given by* $B_\varphi(X,Y) = \nabla^h_{d\varphi(X)} d\varphi(Y) - d\varphi(\nabla^g_X Y)$ *and* $\{e_j\}$ *and* $\{e_j\}$ *is a locally-defined orthonormal frame field on* (M, g).

We obtain:

Theorem 22. *Let* φ: $(M^m, g) \to (N^n, h)$ *be an isometric immersion. Assume that* $\overline{\nabla}^{\perp}_X \tau(\varphi) = 0$ *for all*(X∈X(M)) *Then,* φ *is biharmonic if and only if the following equations hold:*

1. *The tangential part:*

$$\left(\sum_{k=1}^{m} R^h(\tau(\varphi), d\varphi(e_k))d\varphi(e_k) \right)^{\top} = 0 \qquad (85)$$

2. *The normal part:*

$$\left(\sum_{k=1}^{m} R^h(\tau(\varphi), d\varphi(e_k))\, d\varphi(e_k) \right)^{\perp} = \sum_{j,k=1}^{m} h(\tau(\varphi), B_\varphi(e_j, e_k))\, B_\varphi(e_j, e_k)$$

(86)

As a corollary of Theorem 2, we obtain:

Corollary 23. *Assume that the sectional curvature of the target space* (N^n, h) *is non-positive. Let* φ: $(M^m, g) \to (N^n, h)$ *be an isometric immersion whose tension field satisfies* $\overline{\nabla}^{\perp}_X \tau(\varphi) = 0$ *for* vall(X∈X(M)). *Then, if* φ *is biharmonic, then it is harmonic.*

Remark 2. *Corollary 24 gives partial evidence to the generalized B.-Y. Chen's conjecture: every biharmonic isometric immersion into a non-positive curvature manifold must be harmonic. On the other hand, notice that the generalized B.-Y. Chen's conjecture was given by a counter example due to Y. Ou and L. Tang [27].*

7.2. Biharmonic Submanifolds in Einstein Manifolds

In the second part of this section, we apply Theorem 23 to an isometric immersion into an Einstein manifold (N^n, h) whose Ricci transform is denoted by ρ^h, by definition, ρh $\rho^h(u) := \sum_{i=1}^{n} R^h(u, e_i')e_i'$ $(u \in T_yN,\ y \in N)$, where $\{e_i'\}_{i=1}^{n}$ is a locally-defined orthonormal frame field on (N^n, h). Then, we obtain:

Theorem 24. *Assume that* φ: $(M^m, g) \to (N^n, h)$ *is an isometric immersion whose tension field* $\tau(\varphi)$ *that satisfies that* $\overline{\nabla}^{\perp}_X \tau(\varphi) = 0$, *and the target space* (N, h) *is an Einstein, i.e., the Ricci transform* ρ^h *of* (N, h) *satisfies* $\rho^h = c\,\mathrm{Id}$ *for some constant c. Then,* φ *is biharmonic if and only if the following holds:*

$$c\tau(\varphi) - \sum_{i=1}^{p} R^h(\tau(\varphi), \xi_i)\xi_i = \sum_{j,k=1}^{m} h(\tau(\varphi), B_\varphi(e_j, e_k))\, B_\varphi(e_j, e_k)$$

(87)

Where $\{\xi_i\}_{i=1}^{p}$p *is a local orthonormal frame field of the normal bundle corresponding to the immersion* $\varphi: M \to N$.

In the following, we treat with a hypersurface $\varphi: (M^m, g) \to (N^n, h)$, *i.e.*, $p = 1$, and $m = \dim M = \dim N - 1 = n - 1$. In this case, we obtain the following theorem:

Theorem 25. *Assume that* $\varphi: (M^m, g) \to (N^n, h)$ *is an isometric immersion whose tension field*∇¯⊥Xτ(φ)=0(∀X∈X(M))∇¯X⊥τ(φ)=0(∀X∈X(M)) *and* φ *is a hypersurface, i.e.,* $m = n - 1$.

1. *If* φ *is not harmonic, then* φ *is biharmonic if and only if:*

$$\rho^h(\xi) = \|B_\varphi\|^2\, \xi, \quad (88)$$

where ρ^h *is the Ricci transform of* (N, h) *and* ξ *is a unit normal vector field along* φ.

2. *In particular, if* (N, h) *is an Einstein manifold, i.e.,* $\rho^h = c\,\mathrm{Id}$ *and* φ *is not harmonic, then* φ *is biharmonic if and only if* $\|B_\varphi\|^2 = c$.

Furthermore, we have:

Theorem 26. *Assume that* $\varphi: (M, g) \to (N, h)$ *is an isometric immersion into a Riemannian manifold* (N, h) *whose Ricci* *curvature is non-positive,* dim $M = \dim N - 1$, *and* $\overline{\nabla}^{\perp}_{X}\tau(\varphi) = 0$ *for all* C^{∞} *for all* C^{∞} *vector field X on M. Then, if* φ *is biharmonic, it is harmonic.*

Finally in this section, we give a criterion for which the condition $\overline{\nabla}^{\perp}_{X}\tau(\varphi) = 0 \ (\forall X \in$ X(M)) holds:

Proposition 27. *Assume that* φ: $(M^m, g) \rightarrow (N^n, h)$ *is an isometric immersion with* $m = \dim M = \dim N - 1 = n - 1$. *Then, the following equivalence holds: The condition that* $\overline{\nabla}^{\perp}_{X}\tau(\varphi) = 0 \ (\forall X \in$ X(m) *holds if and only if the mean curvature* $\mathbf{H} = \frac{1}{m}\sum_{i=1}^{m} H_{ii}$ *is constant on M. Here,* $B_{\varphi}(e_i, e_j) = H_{ij}\,\xi$, *and* ξ *is a unit normal vector field along* φ.

Summarizing Theorems 26 and 27 and Proposition 28, we obtain:

Corollary 28. *Let* φ: $(M^m, g) \rightarrow (N^n, h)$ *be an isometric immersion. Assume that* $m = \dim M \dim N - 1 = n - 1$ and the mean curvature of φ, $\mathbf{H} = \frac{1}{m}\sum_{i=1}^{m} H_{ii} = \frac{1}{m}\tau(\varphi)$, *is constant. Then, following hold:*

1. *Assume that* $\mathbf{H} \neq 0$, *i.e.,* φ *is not harmonic. Then, it holds that* φ *is biharmonic if and only if* $\rho^h(\xi) = \|B_{\varphi}\|^2\,\xi$, *where* ρ^h *is the Ricci transform of* (N, h), ξ *is a unit normal vector field along* φ *and* B_{φ} *is the second fundamental form of* φ.

2. *Assume that* $\mathbf{H} \neq 0$ *and* (N, h) *is Einstein, i.e.,* $\rho^h = c\,\mathrm{Id}$ *for some constant c. Then,* φ *is biharmonic if and only if* $\|B_{\varphi}\|^2 = c$.

3. *Assume that* $\mathbf{H} \neq 0$ *and the Ricci curvature of* (N, h) *is non-positive. Then, if* φ *biharmonic, it is harmonic.*

Due to the above, we have a classification of homogeneous hypersurfaces in compact symmetric spaces. Refer to our recent paper [32]. Finally, we raise a problem related to this topic (*cf.*, [32]):

Problem 29. *Let* (N^n, h) *be an Einstein manifold with the Ricci transform* $\rho^h = c\,\mathrm{Id}$ *for some constant* $c > 0$ *and admitting a low cohomogeneity action of some compact Lie group* H. *Determine all of the* H*-orbits, which are harmonic or biharmonic.*

8. CONCLUSIONS

We have given a survey on recent progresses on Chen's conjecture and the generalized Chen's conjecture, the outlines of the proofs on the L^2 rigidity theorems of biharmonic maps due to N. Nakauchi and myself, and also its *CR* analogue. Then, we have given a survey on constructions and classification of harmonic maps and biharmonic maps into compact Lie groups or compact symmetric spaces. Finally, we have explained how to construct biharmonic hypersurfaces of compact symmetric spaces. We hope for young geometers to read this survey, and to give new results by extending our results, and to attack and obtain answers of the above problem 29.

CONFLICTS OF INTEREST

The author declares no conflict of interest.

REFERENCES

1. Eells, J.; Lemaire, L. Selected Topics in Harmonic Maps, Proceedings of the CBMS Regional Conference Series in Mathematics, Providence, RI, USA, 31 December 1983.
2. Jiang, G.Y. 2-harmonic maps and their first and second variational formula. *Chinese Ann. Math. A* **1986**, *7*, 388–402.
3. Akutagawa, K.; Maeta, S. Properly immersed biharmonic submanifolds in the Euclidean spaces. *Geom. Dedicata* **2013**,*164*, 351–355.
4. Baird, P.; Fardoun, A.; Ouakkas, S. Liouville-type theorems for biharmonic maps between Riemannian manifolds.*Adv. Calc. Var.* **2010**, *3*, 49–68.
5. Balmus, A.; Montaldo, S.; Oniciuc, C. Classification results for biharmonic submanifolds in spheres. *Isr. J. Math.* **2008**,*168*, 201–220.
6. Balmus, A.; Montaldo, S.; Oniciuc, C. Biharmonic hypersurfaces in 4-dimensional space forms. *Math. Nachr.* **2010**, *283*, 1696–1705.
7. Caddeo, R.; Montaldo, S.; Piu, P. On biharmonic maps. *Contemp. Math.* **2001**, *288*, 286–290.

8. Chen, B.Y. Some open problems and conjectures on submanifolds of finite type. *Soochow J. Math.* **1991**, *17*, 169–188.

9. Defever, F. Hypersurfaces in E4E4 with harmonic mean curvature vector. *Math. Nachr.* **1998**, *196*, 61–69.

10. Dragomir, S.; Montaldo, S. Subelliptic biharmonic maps. *J. Geom. Anal.* **2014**, *24*, 223–245.

11. Fetcu, D.; Oniciuc, C. Biharmonic integral *C*-parallel submanifolds in 7-dimensional Sasakian space forms. *Tohoku Math. J.* **2012**, *64*, 195–222.

12. Hasanis, T.; Vlachos, T. Hypersurfaces in E4E4 with harmonic mean curvaturer vector field. *Math. Nachr.* **1995**, *172*, 145–169.

13. Ichiyama, T.; Inoguchi, J.; Urakawa, H. Biharmonic maps and bi-Yang-Mills fields. *Note Mat.* **2009**, *28*, 233–275.

14. Ichiyama, T.; Inoguchi, J.; Urakawa, H. Classifications and isolation phenomena of biharmonic maps and bi-Yang-Mills fields. *Note Mat.* **2010**, *30*, 15–48.

15. Inoguchi, J. Submanifolds with harmonic mean curvature vector field in contact 3-manifolds. *Colloq. Math.* **2004**, *100*, 163–179.

16. Ishihara, S.; Ishikawa, S. Notes on relatively harmonic immersions. *Hokkaido Math. J.* **1975**, *4*, 234–246.

17. Loubeau, E.; Oniciuc, C. The index of biharmonic maps in spheres. *Compositio Math.* **2005**, *141*, 729–745.

18. Loubeau, E.; Oniciuc, C. On the biharmonic and harmonic indices of the Hopf map. *Trans. Amer. Math. Soc.* **2007**, *359*, 5239–5256.

19. Loubeau, E.; Ou, Y.-L. Biharmonic maps and morphisms from conformal mappings. *Tohoku Math. J.* **2010**, *62*, 55–73.

20. Montaldo, S.; Oniciuc, C. A short survey on biharmonic maps between Riemannian manifolds. *Rev. Un. Mat. Argentina***2006**, *47*, 1–22.

21. Nakauchi, N.; Urakawa, H. Biharmonic hypersurfaces in a Riemannian manifold with non-positive Ricci curvature.*Annu. Global Anal. Geom.* **2011**, *40*, 125–131.

22. Nakauchi, N.; Urakawa, H. Biharmonic submanifolds in a Riemannian manifold with non-positive curvature. *Results Math.* **2013**, *63*, 467–474.

23. Nakauchi, N.; Urakawa, H.; Gudmundsson, S. Biharmonic maps into a Riemannian manifold of non-positive curvature. *Geom. Dedicata* **2014**, *169*, 263–272.

24. Oniciuc, C. Biharmonic maps between Riemannian manifolds. *Annu. Stiint Univ. Aℓ. Cuza Iasi, Mat. (N.S.)* **2002**, *48*, 237–248.

25. Sasahara, T. Legendre surfaces in Sasakian space forms whose mean curvature vectors are eigenvectors. *Publ. Math. Debrecen* **2005**, *67*, 285–303.

26. Sasahara, T. Stability of biharmonic Legendrian submanifolds in Sasakian space forms. *Canad. Math. Bull.* **2008**, *51*, 448–459. [

27. Ou, Y.-L.; Tang, L. *The generalized Chen's conjecture on biharmonic submanifolds is false. 2010, arXiv: 1006.1838v1,* Available online: http://arxiv.org/abs/1006.1838 accessed on 30 April 2015.

28. Wang, Z.-P.; Ou, Y.-L. Biharmonic Riemannian submersions from 3-manifolds. *Math. Z.* **2011**, *269*, 917–925.

29. Urakawa, H. Biharmonic maps into compact Lie groups and integrable systems. *Hokkaido Math. J.* **2014**, *43*, 73–103.

30. Urakawa, H. Biharmonic maps into symmetric spaces and integrable systems. *Hokkaido Math. J.* **2014**, *43*, 105–136.

31. Urakawa, H. *CR* rigidity of pseudo harmonic maps and pseudo biharmonic maps. *Hokkaido Math. J.* In press. 2014. arXiv: 1410.0086v1. Available online: http://arxiv.org/pdf/1410.0086 accessed on 30 April 2015. [

32. Ohno, S.; Sakai, T.; Urakawa, H. *Biharmonic homogeneous hypersurfaces in compact symmetric spaces*; Unpublished work; 2015. [

33. Maeta, S.; Urakawa, U. Biharmonic Lagrangian submanifolds in Kähler manifolds. *Glasgow Math. J.* **2013**, *55*, 465–480.

34. Maeta, S.; Nakauchi, N.; Urakawa, H. *Triharmonic isometric immersions into a manifold of non-positively constant curvature. 2013, arXiv: 1201.6457,* Available

online: http://arxiv.org/abs/1309.0280 accessed on 30 April 2015.

35. Markellos, M.; Urakawa, H. The bienergy of unit vector fields. *Annu. Global Anal. Geom.* **2014**, *46*, 431–457.

36. Markellos, M.; Urakawa, H. *The biharmonicity of sections of the tangent bundle. 2014, arXiv: 1407.1127*, Available online: http://arxiv.org/abs/1407.1127 accessed on 30 April 2015.

37. Montaldo, S.; Ratto, A. Biharmonic submanifolds into ellipsoids. *Monatshefte Math.* **2015**, *176*, 589–601.

38. Naito, H.; Urakawa, H. Conformal change of Riemannian metrics and biharmonic maps. *Indiana Univ. Math. J.* **2014**,*63*, 1631–1657.

39. Nakauchi, N.; Urakawa, H. *Polyharmonic maps into the Euclidean space. 2013, arXiv: 1307.5089v2*, Available online: http://arxiv.org/abs/1307.5089 accessed on 30 April 2015.

40. Urakawa, H. *Sasaki manifolds, Kähler cone manifolds and biharmonic submanifolds. 2013, arXiv: 1306.6123*, Available online:http://arxiv.org/abs/1306.6123 accessed on 30 April 2015.

41. Ou, Y.-L.; Tang, L. On the generalized Chen's conjecture on biharmonic submanifolds. *Michigan Math. J.* **2012**, *61*, 531–542.

42. Sasahara, T. A class of biminimal Legendrian submanifolds in Sasaki space forms. *Math. Nachr.* **2014**, *287*, 79–90.

43. Koiso, N.; Urakawa, H. *Biharmonic submanifolds in a Riemannian manifold. 2014, arXiv: 1408.5494v1*, Available online: http://arxiv.org/abs/1408.5494 accessed on 30 April 2015.

44. Schoen, R.; Yau, S.T. Harmonic maps and the topology of stable hypersurfaces and manifolds with non-negative Ricci curvature. *Comment. Math. Helv.* **1976**, *51*, 333–341.

45. Castro, I.; Li, H.Z.; Urbano, F. Hamiltonian-minimal Lagrangian submanifolds in complex space forms. *Pacific J. Math.***2006**, *227*, 43–63.

46. Kajigaya, T. Second variation formula and the stability of Legendrian minimal submanifolds in Sasakian manifolds.*Diff. Geom. Appl.* **2014**, *37*, 89–108.

47. Nagatomo, Y. Harmonic maps into Grassmannians and a generalization of do Carmo-Wallach theorem, Proceedings of the 16th OCU International Academic Symposium, Osaka, Japan, 15–20 December 2008; pp. 41–52.

48. Takahashi, T. Minimal immersions of Riemannian manifoplds. *J. Math. Soc. Jpn.* **1966**, *18*, 380–385.

49. Iriyeh, H. Hamiltonian minimal Lagrangian cones in $\mathbb{C}^m$. *Tokyo J. Math.* **2005**, *28*, 91–107.

50. Yau, S.T. Some function-theoretic properties of complete Riemannian manifold and their applications to geometry.*Indiana Univ. Math. J.* **1976**, *25*, 659–670.

51. Baird, P.; Eells, J. A conservation law for harmonic maps. *Lect. Notes Math.* **1981**, *894*, 1–25.

52. Baird, P.; Wood, J. *Harmonic Morphisms Between Riemannian Manifolds*; Oxford Science Publication: Oxford, UK, 2003.

53. Gaffney, M.F. A special Stokes' theorem for complete Riemannian manifold. *Annu. Math.* **1954**, *60*, 140–145.

54. Gudmundsson, S. *The Bibliography of Harmonic Morphisms*, Available online: http://matematik.lu.se/matematiklu/personal/sigma/harmonic/bibliography.html accessed on 30 April 2015.

55. Dai, Y.-J.; Shoji, M.; Urakawa, H. Harmonic maps into Lie groups and homogeneous spaces. *Differ. Geom. Appl.* **1997**, *7*, 143–160.

56. Kobayashi, S.; Nomizu, K. *Foundations of Differential Geometry, Volume I*; Interscience Publishers, John Wiley and Sons: New York, NY, USA, 1963.

57. Kobayashi, S.; Nomizu, K. *Foundations of Differential Geometry, Volume II*; Interscience Publishers, John Wiley and Sons: New York, NY, USA, 1969.

58. Chern, S.S.; Moser, J.K. Real hypersurfaces in complex manifolds. *Acta Math.* **1974**, *133*, 48–69.

59. Dragomir, S.; Tomassini, G. *Differential Geometry and Analysis on CR Manifolds*; Birkhouser: Basel, Switzerland, 2006.

60. Barletta, E.; Dragomir, S.; Urakawa, H. Pseudoharmonic maps from a nondegenerate *CR* manifold into a Riemannian manifold. *Indiana Univ. Math. J.* **2001**, *50*, 719–746.

61. Aribi, A. *Le Spectre du Sous-Laplacien sur les Variétés CR Strictement Pseudoconvexes*; Universitè de Tours: Thèse, France, 2012; In French.

10

CHAPTER

Conformal Maps, Biharmonic Maps, and the Warped Product

Seddik Ouakkas * and Djelloul Djebbouri

Laboratory of Geometry, Analysis, Control and Applications, University de Saida, BP138, En-Nasr, 20000 Saida, Algeria

ABSTRACT

In this paper we study some properties of conformal maps between equidimensional manifolds, we construct new example of non-harmonic biharmonic maps and we characterize the biharmonicity of some maps on the warped product manifolds.

Keywords: biharmonic map; conformal map; warped product

1. INTRODUCTION.

Let $(M^m, g) \to (N^n, h)$ be a smooth map between Riemannian manifolds. Then ϕ is said to be harmonic if it is a critical point of the energy functional :

$$E(\phi) = \frac{1}{2}\int_K |d\phi|^2 dv_g \qquad (1)$$

for any compact subset K ⊂ M . Equivalently, ϕ is harmonic if it satisfies the associated Euler-Lagrange equations :

$$\tau(\phi) = Tr_g \nabla d\phi = 0, \qquad (2)$$

and τ(ϕ)τ(ϕ) is called the tension field of ϕ. One can refer to [1,2,3,4] for background on harmonic maps. In the context of harmonic maps, the stress-energy tensor was studied in details by Baird and Eells in [5]. The stress-energy tensor for a map ϕ:(Mm,g)⟶(Nn,h) defined by

$$S(\phi) = e(\phi)g - \phi^* h$$

and the relation between S(ϕ) and τ(ϕ) is given by

$$divS(\phi) = -h(\tau(\phi), d\phi).$$

The map ϕ is said to be biharmonic if it is a critical point of the bi-energy functional :

$$E_2(\phi) = \frac{1}{2} \int_M |\tau(\phi)|^2 dv_g \tag{3}$$

Equivalently, ϕ is biharmonic if it satisfies the associated Euler-Lagrange equations :

$$\tau_2(\phi) = -Tr_g \left(\nabla^\phi\right)^2 \tau(\phi) - Tr_g R^N (\tau(\phi), d\phi) d\phi = 0, \tag{4}$$

where ∇ϕ is the connection in the pull-back bundle ϕ−1(TN) and, if (ei)1≤i≤mei is a local orthonormal frame field on M, then

$$Tr_g \left(\nabla^\phi\right)^2 \tau(\phi) = \left(\nabla^\phi_{e_i} \nabla^\phi_{e_i} - \nabla^\phi_{\nabla_{e_i} e_i}\right) \tau(\phi),$$

where we sum over repeated indices. We will call the operator τ2(ϕ), the bi-tension field of the map ϕ. In analogy with harmonic maps, Jiang In [6] has constructed for a map ϕ the stress bi-energy tensor defined by

$$S_2(\phi) = \left(\frac{-1}{2} |\tau(\phi)|^2 + divh\left(\tau(\phi), d\phi\right)\right) g - 2symh\left(\nabla \tau(\phi), d\phi\right),$$

where

$$symh\left(\nabla \tau(\phi), d\phi\right)(X, Y) = \frac{1}{2} \left\{h\left(\nabla_X \tau(\phi), d\phi(Y)\right) + h\left(\nabla_Y \tau(\phi), d\phi(X)\right)\right\},$$

for any X,Y∈Γ(TM).. The stress bi-energy tensor was also studied in [7] and those results could be useful when we study conformal maps. The stress bi-energy tensor of ϕ satisfies the following relationship

1. Introduction.

$$divS_2(\phi) = h\left(\tau_2(\phi), d\phi\right).$$

Clearly any harmonic map is biharmonic, therefore it is interesting to construct non-harmonic biharmonic maps. In [8] the authors found new examples of biharmonic maps by conformally deforming the domain metric of harmonic ones. While in [9] the author analyzed the behavior of the biharmonic equation under the conformal change the domain metric, she obtained metrics g˜=e2γ such that the idendity map Id:(M,g)⟶(M,g˜) is biharmonic non-harmonic. Moreover, in [10] the author gave some extensions of the result in [9] together with some further constructions of biharmonic maps. The author in [11] deform conformally the codomain metric in order to render a semi-conformal harmonic map biharmonic. In [12] the authors studied the case where ϕ:(Mn,g)⟶(Nn,h) is a conformal mapping between equidimensional manifolds where they show that a conformal mapping ϕ is biharmonic if and only if the gradient of its dilation satisfies a second order elliptic partial differential equation. We can refer the reader to [13], for a survey of biharmonic maps. In the first section of this paper, we present some properties for a conformal mapping ϕ:(Mn,g)⟶(Nn,h), we prove that the stress bi-energy tensor depend only on the dilation (Theorem 1) and we calculate the bitension field of ϕ (Theorem 2). In the last section we study the biharmonicity of some maps on the warped product (Theorem 4 and 5), with this setting we obtain new examples of biharmonic non-harmonic maps.

2. SOME PROPERTIES FOR CONFORMAL MAPS.

We study conformal maps between equidimensional manifolds of the same dimension n≥3n≥3. Note that by a result in [12], any such map can have no critical points and so is a local conformal diffeomorphism. Recall that a mapping ϕ:(Mn,g)→(Nn,h)) is called conformal if there exist a C∞C∞ function λ:M→R+* such that for any X,Y∈Γ(TM) :

$$h(d\phi(X), d\phi(Y)) = \lambda^2 g(X,Y).$$

The function λ is called the dilation for the map ϕ. The tension field and the stress energy tensor for a conformal map are given by (see [14]):

Proposition 1.

Let ϕ: $Let\ \phi : (M^n, g) \to (N^n, h)$ *be a conformal map of dilation λ, we have*

$$(i)\ \ divS(\phi) = (n-2)\lambda^2 d \ln \lambda, \qquad (5)$$

$$(ii)\ \ divh(\tau(\phi), d\phi) = (2-n)\left(2\lambda^2 \left|grad \ln \lambda\right|^2 + \lambda^2 \Delta \ln \lambda\right) \qquad (6)$$

$$(iii)\ \ \tau(\phi) = (2-n) d\phi(grad \ln \lambda). \qquad (7)$$

$$(iv)\ \ \left|\tau(\phi)\right|^2 = (2-n)^2 \lambda^2 \left|grad \ln \lambda\right|^2. \qquad (8)$$

Note that the conformal map $\phi:(M^n,g)\to(N^n,h)$ *of dilation* λ *is harmonic if and only if* n=2n=2 *or the dilation* λ *is constant.*

In the first, wa calculate the stress bi-energy tensor for a conformal map ϕ when we prove that $S_2(\phi)$ depend only the dilation.

Theorem 1.

Let $\phi:(M^n,g)\to(N^n,h)$ *be a conformal map with dilation* λ*, then we have*

$$S_2(\phi) = (2-n)\,\lambda^2\left\{\left(\frac{n-2}{2}\,|grad\ln\lambda|^2+\Delta\ln\lambda\right)g-2\nabla d\ln\lambda\right\}, \tag{9}$$

and the trace of $S_2(\phi)$ *is given by*

$$TrS_2(\phi) = -\,(2-n)^2\,\lambda^2\left\{\frac{n}{2}\,|grad\ln\lambda|^2+\Delta\ln\lambda\right\}. \tag{10}$$

To prove Theorem 1, we need the following Lemma:

Lemma 1.

Let $\phi:(M^n,g)\to(N^n,h)$ *be a conformal map with dilation* λ*, then for any function* $f\in C^\infty(M)$ *and for any* $X,Y\in\Gamma(TM)$*, we have*

$$\begin{aligned} h\left(\nabla_X d\phi\,(gradf)\,,d\phi\,(Y)\right) &= \lambda^2\left(X\,(\ln\lambda)\,Y\,(f)-Y\,(\ln\lambda)\,X\,(f)\right)\\ &\quad+\lambda^2\nabla df\,(X,Y)+\lambda^2 d\ln\lambda\,(gradf)\,g\,(X,Y). \end{aligned} \tag{11}$$

Proof of Lemma 1. Let $f\in C^\infty(M)$, for any $X,Y\in\Gamma TM$, we have

$$\begin{aligned} h\left(\nabla_X d\phi\left(gradf\right), d\phi\left(Y\right)\right) &= X\left(\lambda^2 g\left(gradf, Y\right)\right) - h\left(d\phi\left(gradf\right), \nabla_X d\phi\left(Y\right)\right) \\ &= X\left(\lambda^2\right) g\left(gradf, Y\right) + \lambda^2 g\left(\nabla_X gradf, Y\right) + \lambda^2 g\left(gradf, \nabla_X Y\right) \\ &- h\left(d\phi\left(gradf\right), \nabla d\phi\left(X, Y\right)\right) - h\left(d\phi\left(gradf\right), d\phi\left(\nabla_X Y\right)\right) \\ &= X\left(\lambda^2\right) g\left(gradf, Y\right) + \lambda^2 g\left(\nabla_X gradf, Y\right) + \lambda^2 g\left(gradf, \nabla_X Y\right) \\ &- h\left(d\phi\left(gradf\right), \nabla d\phi\left(X, Y\right)\right) - \lambda^2 g\left(gradf, \nabla_X Y\right). \end{aligned}$$

Note that

$$g\left(\nabla_X gradf, Y\right) = \nabla df\left(X, Y\right),$$

then we obtain

$$h\left(\nabla_X d\phi\left(gradf\right), d\phi\left(Y\right)\right) = 2\lambda^2 X\left(\ln\lambda\right) Y\left(f\right) + \lambda^2 \nabla df\left(X, Y\right) - h\left(d\phi\left(gradf\right), \nabla d\phi\left(X, Y\right)\right).$$

By similary, we have

$$h\left(\nabla_Y d\phi\left(gradf\right), d\phi\left(X\right)\right) = 2\lambda^2 Y\left(\ln\lambda\right) X\left(f\right) + \lambda^2 \nabla df\left(X, Y\right) - h\left(d\phi\left(gradf\right), \nabla d\phi\left(X, Y\right)\right).$$

Then, we deduce that

$$\begin{aligned} h\left(\nabla_X d\phi\left(gradf\right), d\phi\left(Y\right)\right) &= h\left(d\phi\left(X\right), \nabla_Y d\phi\left(gradf\right)\right) \\ &+ 2\lambda^2\left(X\left(\ln\lambda\right) Y\left(f\right) - Y\left(\ln\lambda\right) X\left(f\right)\right). \end{aligned} \tag{12}$$

For the term $h\left(d\phi\left(X\right), \nabla_Y d\phi\left(gradf\right)\right),$ we have

$$\begin{aligned} h\left(\nabla_Y d\phi\left(gradf\right), d\phi\left(X\right)\right) &= h\left(\nabla d\phi\left(gradf, Y\right), d\phi\left(X\right)\right) + \lambda^2 g\left(\nabla_Y gradf, X\right) \\ &= h\left(\nabla_{gradf} d\phi\left(Y\right), d\phi\left(X\right)\right) - \lambda^2 g\left(\nabla_{gradf} Y, X\right) \\ &+ \lambda^2 g\left(\nabla_Y gradf, X\right) \\ &= gradf\left(\lambda^2 g\left(X, Y\right)\right) - h\left(\nabla_{gradf} d\phi\left(X\right), d\phi\left(Y\right)\right) \\ &- \lambda^2 g\left(\nabla_{gradf} Y, X\right) + \lambda^2 g\left(\nabla_Y gradf, X\right) \\ &= 2\lambda^2 d\ln\lambda\left(gradf\right) g\left(X, Y\right) - h\left(\nabla d\phi\left(X, gradf\right), d\phi\left(Y\right)\right) \\ &+ \lambda^2 g\left(\nabla_Y gradf, X\right). \end{aligned}$$

We deduce that

$$\begin{aligned} h\left(\nabla_Y d\phi\left(gradf\right), d\phi\left(X\right)\right) = -h\left(\nabla_X d\phi\left(gradf\right), d\phi\left(Y\right)\right) + 2\lambda^2 \nabla df\left(X, Y\right) \\ + 2\lambda^2 d\ln\lambda\left(gradf\right) g\left(X, Y\right). \end{aligned} \tag{13}$$

Finally, if we replace (13) in (12), we obtain

$$\begin{aligned} h\left(\nabla_X d\phi\left(gradf\right), d\phi\left(Y\right)\right) = \lambda^2\left(X\left(\ln\lambda\right) Y\left(f\right) - Y\left(\ln\lambda\right) X\left(f\right)\right) \\ + \lambda^2 \nabla df\left(X, Y\right) + \lambda^2 d\ln\lambda\left(gradf\right) g\left(X, Y\right). \end{aligned}$$

This completes the proof of Lemma 1.

Remark 1.

Let ϕ:(Mn,g)→(Nn,h) *be a conformal map with dilation λ, then if we consider* f=lnλ *the equation (11) gives*

$$h\left(\nabla_X d\phi\left(grad\ln\lambda\right), d\phi\left(Y\right)\right) = \lambda^2\left(\nabla d\ln\lambda\left(X, Y\right) + \left|grad\ln\lambda\right|^2 g\left(X, Y\right)\right).$$

Proof of Theorem 1. By definition, the stress bi-energy tensor is given by

$$S_2(\phi) = \left(-\frac{1}{2}|\tau(\phi)|^2 + divh(\tau(\phi), d\phi)\right) g - 2symh(\nabla\tau(\phi), d\phi). \tag{14}$$

Using the equations (2) et (4) for the Proposition 1, we have

$$-\frac{1}{2}|\tau(\phi)|^2 + divh(\tau(\phi), d\phi) = (2-n)\lambda^2 \left(\frac{n+2}{2}|grad \ln\lambda|^2 + \Delta \ln\lambda\right). \tag{15}$$

Calculate now symh(∇τ(ϕ),dϕ), we have by definition for any X,Y∈Γ(TM)

$$\begin{aligned} symh(\nabla\tau(\phi), d\phi)(X,Y) &= \frac{1}{2}\left(h(\nabla_X\tau(\phi), d\phi(Y)) + h(\nabla_Y\tau(\phi), d\phi(X))\right) \\ &= \frac{2-n}{2}\left(h(\nabla_X d\phi(grad \ln\lambda), d\phi(Y)) + h(\nabla_Y(grad \ln\lambda), d\phi(X))\right). \end{aligned}$$

By Lemma 1, we have

$$h(\nabla_X d\phi(grad \ln\lambda), d\phi(Y)) = \lambda^2\left(\nabla d\ln\lambda(X,Y) + |grad \ln\lambda|^2 g(X,Y)\right)$$

and

$h(\nabla_Y d\phi(grad \ln\lambda), d\phi(X)) = \lambda^2\left(\nabla d\ln\lambda(X,Y) + |grad \ln\lambda|^2 g(X,Y)\right)$, then

$$symh(\nabla\tau(\phi), d\phi)(X,Y) = (2-n)\lambda^2\left(\nabla d\ln\lambda(X,Y) + |grad \ln\lambda|^2 g(X,Y)\right). \tag{16}$$

If we substitute (15) and (16) in (14), we conclude that

$$S_2(\phi) = (2-n)\lambda^2\left\{\left(\frac{n-2}{2}|grad \ln\lambda|^2 + \Delta\ln\lambda\right) g - 2\nabla d\ln\lambda\right\}$$

Calculate now the trace of stress bi-energy tensor. Let $(e_i)_{1\leq i\leq n}$ an orthonormal frame on *M*, we have

$$\begin{aligned}
Tr_g S_2(\phi) &= S_2(\phi)(e_i, e_i) \\
&= (2-n)\,\lambda^2 \left(\frac{n-2}{2}\,|grad\ln\lambda|^2 + \Delta\ln\lambda\right) g\,(e_i, e_i) \\
&- 2\,(2-n)\,\lambda^2 \nabla d\ln\lambda\,(e_i, e_i) \\
&= (2-n)\,n\lambda^2 \left(\frac{n-2}{2}\,|grad\ln\lambda|^2 + \Delta\ln\lambda\right) \\
&- 2\,(2-n)\,\lambda^2\,(\Delta\ln\lambda) \\
&= (2-n)\,\lambda^2 \left\{\frac{n\,(n-2)}{2}\,|grad\ln\lambda|^2 + (n-2)\,\Delta\ln\lambda\right\}.
\end{aligned}$$

Then

$$TrS_2(\phi) = -\,(2-n)^2\,\lambda^2 \left\{\frac{n}{2}\,|grad\ln\lambda|^2 + \Delta\ln\lambda\right\}.$$

By calculating the Laplacian of the function λn2 and by using

$$\Delta\lambda^{\frac{n}{2}} = \frac{n}{2}\lambda^{\frac{n}{2}} \left(\frac{n}{2}\,|grad\ln\lambda|^2 + \Delta\ln\lambda\right),$$

we obtain immediately the following corollary

Corollary 1.

Let ϕ:(Mn,g)→(Nn,h)ϕ: $(M^n, g) \to (N^n, h)$, $(n \neq 2)$ *to be a conformal map of dilation λ, then the trace of* S2(ϕ) *is zero if and only if the function* λn2λn2 *is harmonic.*

The bi-tension field of the conformal map is given by

Theorem 2.

Let ϕ:(Mn,g)→(Nn,h) *to be a conformal map of dilation λ, then bi-tension field of ϕ is given by :*

$$\tau_2(\phi) = (n-2)\, d\phi\,(H)$$

where

$$H = grad\Delta \ln \lambda - \frac{(n-6)}{2} grad\left(|grad \ln \lambda|^2\right) + 2Ricci^M\,(grad \ln \lambda)$$
$$-\left(2\,(\Delta \ln \lambda) + (n-2)\,|grad \ln \lambda|^2\right) grad \ln \lambda. \tag{17}$$

Remark 2.

A. Balmus in [9] studied the case where ϕ=IdM, *she obtained the biharmonicity of the identity map from* (M,g)M,g *onto*(M,λ2g)M,λ2g, *this case was also studied in [15].*

To prove the Theorem 2, we need two Lemmas. In the first Lemma, we give a simple formula of the term $Tr_g\,(\nabla^\phi)^2\, d\phi\,(grad\gamma)$ for a conformal map $\phi : (M^n, g) \rightarrow (N^n, h)$ $(n \geq 3)$ of dilation λ and for any function $\gamma \in C^\infty(M)$.

Lemma 2.

Let ϕ:(Mn,g)→(Nn,h) *to be a conformal map of dilation* λ, *then for any function* $\gamma \in C^\infty(M)$. *, we have*

$$Tr_g\,(\nabla^\phi)^2\, d\phi\,(grad\gamma) = d\phi\,(grad\Delta\gamma) + 4d\phi\left(\nabla_{grad \ln \lambda} grad\gamma\right) + d\phi\left(Ricci^M\,(grad\gamma)\right)$$
$$+ (\Delta \ln \lambda)\, d\phi\,(grad\gamma) - 2\,(\Delta\gamma)\, d\phi\,(grad \ln \lambda)$$
$$-(n-2)\, d\ln\lambda\,(grad\gamma)\, d\phi\,(grad \ln \lambda). \tag{18}$$

Proof of Lemma 2. Let γ∈C∞(M), by definition, we have

$$Tr_g\left(\nabla^{\phi}\right)^2 d\phi\left(grad\gamma\right)=\nabla^{\phi}_{e_i}\nabla^{\phi}_{e_i}d\phi\left(grad\gamma\right)-\nabla^{\phi}_{\nabla_{e_i}e_i}d\phi\left(grad\gamma\right).$$

(19)

(Here henceforth we sum over repeated indices.) Let us start with the calculation of the term $\nabla^{\phi}_{e_i}\nabla^{\phi}_{e_i}d\phi\left(grad\gamma\right)$, we have

$$\nabla^{\phi}_{e_i}d\phi\left(grad\gamma\right)=\nabla d\phi\left(e_i,grad\gamma\right)+d\phi\left(\nabla_{e_i}grad\gamma\right).$$

It is known that (see [16])

$$\nabla d\phi\left(e_i,grad\gamma\right)=e_i\left(\ln\lambda\right)d\phi\left(grad\gamma\right)+d\ln\lambda\left(grad\gamma\right)d\phi\left(e_i\right)-e_i\left(\gamma\right)d\phi\left(grad\ln\lambda\right),$$

then

$$\begin{aligned}\nabla^{\phi}_{e_i}d\phi\left(grad\gamma\right)=&\ e_i\left(\ln\lambda\right)d\phi\left(grad\gamma\right)+d\ln\lambda\left(grad\gamma\right)d\phi\left(e_i\right)\\&-e_i\left(\gamma\right)d\phi\left(grad\ln\lambda\right)+d\phi\left(\nabla_{e_i}grad\gamma\right).\end{aligned}$$

(20)

It follows that

$$\begin{aligned}\nabla^{\phi}_{e_i}\nabla^{\phi}_{e_i}d\phi\left(grad\gamma\right)=&\ \nabla^{\phi}_{e_i}\left\{e_i\left(\ln\lambda\right)d\phi\left(grad\gamma\right)\right\}+\nabla^{\phi}_{e_i}\left\{d\ln\lambda\left(grad\gamma\right)d\phi\left(e_i\right)\right\}\\&-\nabla^{\phi}_{e_i}\left\{e_i\left(\gamma\right)d\phi\left(grad\ln\lambda\right)\right\}+\nabla^{\phi}_{e_i}d\phi\left(\nabla_{e_i}grad\gamma\right).\end{aligned}$$

(21)

We will study term by term the right-hand of this expression. For the first term $\nabla^{\phi}_{e_i}\left\{e_i\left(\ln\lambda\right)d\phi\left(grad\gamma\right)\right\}$, we have

$$\nabla^{\phi}_{e_i}\left\{e_i\left(\ln\lambda\right)d\phi\left(grad\gamma\right)\right\}=e_i\left(\ln\lambda\right)\nabla^{\phi}_{e_i}d\phi\left(grad\gamma\right)+e_i\left(e_i\left(\ln\lambda\right)\right)d\phi\left(grad\gamma\right).$$

By using the Equation (20), we deduce that

$$\begin{aligned}\nabla^{\phi}_{e_i}\{e_i(\ln\lambda)\,d\phi(grad\gamma)\} &= e_i(\ln\lambda)\,e_i(\ln\lambda)\,d\phi(grad\gamma)+e_i(\ln\lambda)\,d\ln\lambda(grad\gamma)\,d\phi(e_i)\\ &-e_i(\ln\lambda)\,e_i(\gamma)\,d\phi(grad\ln\lambda)+e_i(\ln\lambda)\,d\phi(\nabla_{e_i}grad\gamma)\\ &+e_i(e_i(\ln\lambda))\,d\phi(grad\gamma),\end{aligned}$$

then, we obtain

$$\begin{aligned}\nabla^{\phi}_{e_i}\{e_i(\ln\lambda)\,d\phi(grad\gamma)\} &= |grad\ln\lambda|^2\,d\phi(grad\gamma)+d\phi\left(\nabla_{grad\ln\lambda}grad\gamma\right)\\ &+e_i(e_i(\ln\lambda))\,d\phi(grad\gamma).\end{aligned} \tag{22}$$

For the second term $\nabla^{\phi}_{e_i}\{d\ln\lambda(grad\gamma)\,d\phi(e_i)\}$, a similar calculation gives

$$\begin{aligned}\nabla^{\phi}_{e_i}\{d\ln\lambda(grad\gamma)\,d\phi(e_i)\} &= d\ln\lambda(grad\gamma)\,\nabla^{\phi}_{e_i}d\phi(e_i)+e_i\{g(grad\ln\lambda,grad\gamma)\}\,d\phi(e_i)\\ &= d\ln\lambda(grad\gamma)\,\nabla^{\phi}_{e_i}d\phi(e_i)+g(\nabla_{e_i}grad\ln\lambda,grad\gamma)\,d\phi(e_i)\\ &+g(grad\ln\lambda,\nabla_{e_i}grad\gamma)\,d\phi(e_i)\\ &= d\ln\lambda(grad\gamma)\,\nabla^{\phi}_{e_i}d\phi(e_i)+g\left(\nabla_{grad\gamma}grad\ln\lambda,e_i\right)d\phi(e_i)\\ &+g\left(\nabla_{grad\ln\lambda}grad\gamma,e_i\right)d\phi(e_i),\end{aligned}$$

it follows that

$$\begin{aligned}\nabla^{\phi}_{e_i}\{d\ln\lambda(grad\gamma)\,d\phi(e_i)\} &= d\ln\lambda(grad\gamma)\,\nabla^{\phi}_{e_i}d\phi(e_i)+d\phi\left(\nabla_{grad\gamma}grad\ln\lambda\right)\\ &+d\phi\left(\nabla_{grad\ln\lambda}grad\gamma\right).\end{aligned} \tag{23}$$

For the third term $\nabla^{\phi}_{e_i}\{e_i(\gamma)\,d\phi(grad\ln\lambda)\}$, by using the same calculation method and the equation (20), we have

$$\begin{aligned}\nabla^{\phi}_{e_i}\{e_i(\gamma)\,d\phi(grad\ln\lambda)\} &= e_i(\gamma)\,\nabla^{\phi}_{e_i}d\phi(grad\ln\lambda) + e_i(e_i(\gamma))\,d\phi(grad\ln\lambda)\\ &= e_i(\gamma)\,e_i(\ln\lambda)\,d\phi(grad\ln\lambda) + e_i(\gamma)\,d\ln\lambda(grad\ln\lambda)\,d\phi(e_i)\\ &- e_i(\gamma)\,e_i(\ln\lambda)\,d\phi(grad\ln\lambda) + e_i(\gamma)\,d\phi(\nabla_{e_i}grad\ln\lambda)\\ &+ e_i(e_i(\gamma))\,d\phi(grad\ln\lambda),\end{aligned}$$

which gives us

$$\begin{aligned}\nabla^{\phi}_{e_i}\{e_i(\gamma)\,d\phi(grad\ln\lambda)\} &= |grad\ln\lambda|^2\,d\phi(grad\gamma) + d\phi\left(\nabla_{grad\gamma}grad\ln\lambda\right)\\ &+ e_i(e_i(\gamma))\,d\phi(grad\ln\lambda).\end{aligned} \tag{24}$$

Now let us look at the last term $\nabla^{\phi}_{e_i}d\phi(\nabla_{e_i}grad\gamma)$, eigradγ, a simple calculation gives

$$\begin{aligned}\nabla^{\phi}_{e_i}d\phi(\nabla_{e_i}grad\gamma) &= e_i(\ln\lambda)\,d\phi(\nabla_{e_i}grad\gamma) + d\ln\lambda(\nabla_{e_i}grad\gamma)\,d\phi(e_i)\\ &- g(e_i,\nabla_{e_i}grad\gamma)\,d\phi(grad\ln\lambda) + d\phi(\nabla_{e_i}\nabla_{e_i}grad\gamma)\\ &= 2d\phi\left(\nabla_{grad\ln\lambda}grad\gamma\right) - (\Delta\gamma)\,d\phi(grad\ln\lambda)\\ &+ d\phi(\nabla_{e_i}\nabla_{e_i}grad\gamma),\end{aligned}$$

then

$$\begin{aligned}\nabla^{\phi}_{e_i}d\phi(\nabla_{e_i}grad\gamma) &= d\phi(\nabla_{e_i}\nabla_{e_i}grad\gamma) + 2d\phi\left(\nabla_{grad\ln\lambda}grad\gamma\right)\\ &- (\Delta\gamma)\,d\phi(grad\ln\lambda).\end{aligned} \tag{25}$$

If we replace (22), (23), (24) and (25) in (21), we obtain

$$\nabla^{\phi}_{e_i}\nabla^{\phi}_{e_i}d\phi\left(grad\gamma\right) = 4d\phi\left(\nabla_{grad\ln\lambda}grad\gamma\right) + e_i\left(e_i\left(\ln\lambda\right)\right)d\phi\left(grad\gamma\right)$$
$$+ d\ln\lambda\left(grad\gamma\right)\nabla^{\phi}_{e_i}d\phi\left(e_i\right) - e_i\left(e_i\left(\gamma\right)\right)d\phi\left(grad\ln\lambda\right)$$
$$+ d\phi\left(\nabla_{e_i}\nabla_{e_i}grad\gamma\right) - \left(\Delta\gamma\right)d\phi\left(grad\ln\lambda\right). \tag{26}$$

To complete the proof, it remains to investigate the term $\nabla^{\phi}_{\nabla_{e_i}e_i}d\phi\left(grad\gamma\right)$, we have

$$\nabla^{\phi}_{\nabla_{e_i}e_i}d\phi\left(grad\gamma\right) = \nabla d\phi\left(\nabla_{e_i}e_i, grad\gamma\right) + d\phi\left(\nabla_{\nabla_{e_i}e_i}grad\gamma\right),$$

Therefore, by using the equation (20), we obtain

$$\nabla^{\phi}_{\nabla_{e_i}e_i}d\phi\left(grad\gamma\right) = \nabla_{e_i}e_i\left(\ln\lambda\right)d\phi\left(grad\gamma\right) + d\ln\lambda\left(grad\gamma\right)d\phi\left(\nabla_{e_i}e_i\right)$$
$$- \nabla_{e_i}e_i\left(\gamma\right)d\phi\left(grad\ln\lambda\right) + d\phi\left(\nabla_{\nabla_{e_i}e_i}grad\gamma\right). \tag{27}$$

By substituting (26) and (27) in (19), we deduce

$$Tr_g\left(\nabla^{\phi}\right)^2 d\phi\left(grad\gamma\right) = \nabla^{\phi}_{e_i}\nabla^{\phi}_{e_i}d\phi\left(grad\gamma\right) - \nabla^{\phi}_{\nabla_{e_i}e_i}d\phi\left(grad\gamma\right)$$
$$= d\phi\left(Tr_g\nabla^2 grad\gamma\right) + 4d\phi\left(\nabla_{grad\ln\lambda}grad\gamma\right)$$
$$+ \left(\Delta\ln\lambda\right)d\phi\left(grad\gamma\right) + d\ln\lambda\left(grad\gamma\right)\tau\left(\phi\right)$$
$$- 2\left(\Delta\gamma\right)d\phi\left(grad\ln\lambda\right).$$

Finally, using the fact that (see [11])

$$Tr_g\nabla^2 grad\gamma = grad\Delta\gamma + Ricci^M\left(grad\gamma\right)$$

and

$$\tau(\phi) = (2-n)\, d\phi\,(grad \ln \lambda),$$

we conclude that

$$\begin{aligned} Tr_g (\nabla^{\phi})^2 d\phi\,(grad\gamma) &= d\phi\,(grad\Delta\gamma) + 4d\phi\left(\nabla_{grad \ln \lambda} grad\gamma\right) + d\phi\left(Ricci^M (grad\gamma)\right) \\ &+ (\Delta \ln \lambda)\, d\phi\,(grad\gamma) - 2\,(\Delta\gamma)\, d\phi\,(grad \ln \lambda) \\ &- (n-2)\, d \ln \lambda\,(grad\gamma)\, d\phi\,(grad \ln \lambda)\,. \end{aligned}$$

This completes the proof of Lemma 2. Now, in the second Lemma, we will calculate $Tr_g R^N (d\phi\,(grad\gamma), d\phi)\, d\phi$ for a conformal maps ϕ:(Mn,g)→(Nn,h) (n≥3n≥3) of dilation λ and for any function γ∈C∞(M)γ∈C∞M

Lemma 3.

Let ϕ:(Mn,g)→(Nn,h)ϕ: *to be a conformal map of dilation λ, then for any function* γ∈C∞(M), *we have*

$$\begin{aligned} Tr_g R^N (d\phi\,(grad\gamma), d\phi)\, d\phi &= d\phi\left(Ricci^M (grad\gamma)\right) - (n-2)\, d\phi\left(\nabla_{grad\gamma} grad \ln \lambda\right) \\ &- \left(\Delta \ln \lambda + (n-2)\, |grad \ln \lambda|^2\right) d\phi\,(grad\gamma) \\ &+ (n-2)\, d \ln \lambda\,(grad\gamma)\, d\phi\,(grad \ln \lambda) \end{aligned}$$

(28)

Proof of Lemma 3. Let γ∈C∞(M), by definition we have

$$Tr_g R^N (d\phi\,(grad\gamma), d\phi)\, d\phi = R^N (d\phi\,(grad\gamma), d\phi\,(e_i))\, d\phi\,(e_i)$$

(29)

but we know that (see [16])

$$\begin{aligned} Ric^N(d\phi(X), d\phi(Y)) &= Ric^M(X,Y) + (n-2)X(\ln\lambda)Y(\ln\lambda) \\ &- (n-2)|grad \ln\lambda|^2 g(X,Y) \\ &- (n-2)\nabla d\ln\lambda(X,Y) - (\Delta\ln\lambda)g(X,Y). \end{aligned}$$

Then

$$\begin{aligned} Ric^N(d\phi(grad\gamma), d\phi(e_i)) &= Ric^M(grad\gamma, e_i) + (n-2)grad\gamma(\ln\lambda)e_i(\ln\lambda) \\ &- (n-2)|grad\ln\lambda|^2 g(grad\gamma, e_i) \\ &- (n-2)\nabla d\ln\lambda(grad\gamma, e_i) - (\Delta\ln\lambda)g(grad\gamma, e_i) \end{aligned}$$

it follows that

$$\begin{aligned} Ric^N(d\phi(grad\gamma), d\phi(e_i)) &= Ric^M(grad\gamma, e_i) + (m-2)d\ln\lambda(grad\gamma)e_i(\ln\lambda) \\ &- (n-2)|grad\ln\lambda|^2 e_i(\gamma) - (n-2)\nabla d\ln\lambda(grad\gamma, e_i) \\ &- (\Delta\ln\lambda)e_i(\gamma). \end{aligned} \tag{30}$$

If we replace (30) in (29), we deduce that

$$\begin{aligned} Tr_g R^N(d\phi(grad\gamma), d\phi)d\phi &= R^N(d\phi(grad\gamma), d\phi(e_i))d\phi(e_i) \\ &= d\phi\left(Ricci^M(grad\gamma)\right) + (n-2)d\ln\lambda(grad\gamma)d\phi(grad\ln\lambda) \\ &- (n-2)|grad\ln\lambda|^2 d\phi(grad\gamma) - (n-2)\nabla d\ln\lambda(grad\gamma, e_i)d\phi(e_i) \\ &- (\Delta\ln\lambda)d\phi(grad\gamma). \end{aligned}$$

To complete the proof, we will simplify the term ∇dlnλ(gradγ,ei)dϕ(ei), we obtain

$$\begin{aligned} \nabla d\ln\lambda(grad\gamma, e_i)d\phi(e_i) &= \{e_i(g(grad\ln\lambda, grad\gamma)) - d\ln\lambda(\nabla_{e_i}grad\gamma)\}d\phi(e_i) \\ &= g(\nabla_{e_i}grad\ln\lambda, grad\gamma)d\phi(e_i) \\ &= g\left(\nabla_{grad\gamma}grad\ln\lambda, e_i\right)d\phi(e_i) \\ &= d\phi\left(\nabla_{grad\gamma}grad\ln\lambda\right), \end{aligned}$$

which finally gives

$$Tr_g R^N \left(d\phi \left(grad\gamma\right), d\phi\right) d\phi = d\phi \left(Ricci^M \left(grad\gamma\right)\right) - (n-2) d\phi \left(\nabla_{grad\gamma} grad \ln \lambda\right)$$
$$- \left(\Delta \ln \lambda + (n-2) \left|grad \ln \lambda\right|^2\right) d\phi \left(grad\gamma\right)$$
$$+ (n-2) d \ln \lambda \left(grad\gamma\right) d\phi \left(grad \ln \lambda\right).$$

This completes the proof of Lemma 3. We are now able to prove Theorem 2.

Proof of Theorem 2. By definition, the bi-tension field is given by

$$\tau_2 (\phi) = -Tr_g \left(\nabla^\phi\right)^2 \tau (\phi) - Tr_g R^N \left(\tau (\phi), d\phi\right) d\phi.$$

The tension field of the conformal map ϕ is given by

$$\tau (\phi) = (2-n) d\phi \left(grad \ln \lambda\right),$$

it follows that

$$\tau_2 (\phi) = (n-2) \left(Tr_g \left(\nabla^\phi\right)^2 d\phi \left(grad \ln \lambda\right) + Tr_g R^N \left(d\phi \left(grad \ln \lambda\right), d\phi\right) d\phi\right). \tag{31}$$

By Lemma 2, we have

$$Tr_g \left(\nabla^\phi\right)^2 d\phi \left(grad \ln \lambda\right) = d\phi \left(grad\Delta \ln \lambda\right) + 2d\phi \left(grad \left(\left|grad \ln \lambda\right|^2\right)\right)$$
$$- \left(\Delta \ln \lambda\right) d\phi \left(grad \ln \lambda\right) + d\phi \left(Ricci^M \left(grad \ln \lambda\right)\right)$$
$$- (n-2) \left|grad \ln \lambda\right|^2 d\phi \left(grad \ln \lambda\right). \tag{32}$$

By using lemma 3 and the fact that $\nabla_{grad \ln \lambda} grad \ln \lambda = \frac{1}{2} grad \left(\left|grad \ln \lambda\right|^2\right)$

$$Tr_g R^N\left(d\phi\left(grad\ln\lambda\right), d\phi\right)d\phi = d\phi\left(Ricci^M\left(grad\ln\lambda\right)\right) - \left(\Delta\ln\lambda\right)d\phi\left(grad\ln\lambda\right) - \frac{(n-2)}{2}d\phi\left(grad\left(|grad\ln\lambda|^2\right)\right). \tag{33}$$

If we replace (32) and (33) in (31), we deduce that

$$\begin{aligned}\tau_2(\phi) &= (n-2)\,d\phi\left(grad\Delta\ln\lambda\right) - \frac{(n-2)(n-6)}{2}d\phi\left(grad\left(|grad\ln\lambda|^2\right)\right)\\ &- (n-2)\left(2\left(\Delta\ln\lambda\right) + (n-2)\,|grad\ln\lambda|^2\right)d\phi\left(grad\ln\lambda\right)\\ &+ 2(n-2)\,d\phi\left(Ricci^M\left(grad\ln\lambda\right)\right).\end{aligned}$$

Then the bi-tension field of ϕ is given by :

$$\tau_2(\phi) = (n-2)\,d\phi\,(H)$$

where

$$\begin{aligned}H &= grad\Delta\ln\lambda - \frac{(n-6)}{2}grad\left(|grad\ln\lambda|^2\right) + 2Ricci^M\left(grad\ln\lambda\right)\\ &- \left(2\left(\Delta\ln\lambda\right) + (n-2)\,|grad\ln\lambda|^2\right)grad\ln\lambda.\end{aligned}$$

The proof of Theorem 2 is complete. By application of Theorem 2, we get the following result (see [15]).

Theorem 3.

([12]) Let ϕ:(Mn,g)→(Nn,h) *to be a conformal map of dilation* λ*, then* ϕ *is biharmonic if and only if the dilation* λ *satisfies*

$$\begin{aligned}&grad\left(\Delta\ln\lambda\right) - \left(2\left(\Delta\ln\lambda\right) + (n-2)\,|grad\ln\lambda|^2\right)grad\ln\lambda\\ &+ \frac{6-n}{2}grad\left(|grad\ln\lambda|^2\right) + 2Ricci^M\left(grad\ln\lambda\right) = 0.\end{aligned}$$

In particular, we prove that the biharmonicity of the conformal map $\phi:(\mathbb{R}^n,g)\rightarrow(N^n,h)\ (n\geq 3)$ where the dilation $\lambda$is radial $(\ln\lambda=\alpha(r), r=|x| \text{ and } \alpha\in C^\infty(\mathbb{R},\mathbb{R}))$ is equivalent to an ordinary differential equation of the second order. More precisely, we have

Corollary 2.

Let ϕ: $(\mathbb{R}^n,g)\rightarrow(N^n,h)\ (n\geq 3)$ *to be a conformal map of dilation* λ *when we suppose that* lnλ *is radial* $(\ln\lambda=\alpha(r), r=|x| \textit{ and } \alpha\in C^\infty(\mathbb{R},\mathbb{R}))$. *Then* ϕ *is biharmonic if and only if* β=α′β=α′ *satisfies the following ordinary differential equation :*

$$\beta''-(n-4)\beta\beta'+\frac{n-1}{r}\beta'-\frac{n-1}{r^2}\beta-\frac{2(n-1)}{r}\beta^2-(n-2)\beta^3=0. \tag{34}$$

Proof of Corollary 2 Let ϕ: $(\mathbb{R}^n,g)\rightarrow(N^n,h)\ (n\geq 3)$ to be a conformal map of dilation λ such that lnλ=α(r). By Theorem 3, ϕ is biharmonic if and only if the dilation λ satisfies

$$grad(\Delta\ln\lambda)-\left(2(\Delta\ln\lambda)+(n-2)|grad\ln\lambda|^2\right)grad\ln\lambda$$
$$+\frac{6-n}{2}grad\left(|grad\ln\lambda|^2\right)+2Ricci^M(grad\ln\lambda)=0.$$

A direct calculation gives

$$grad\ln\lambda=\alpha'\frac{\partial}{\partial r},$$

$$|grad \ln \lambda|^2 = (\alpha')^2,$$

$$grad\left(|grad \ln \lambda|^2\right) = 2\alpha'\alpha''\frac{\partial}{\partial r},$$

$$\Delta \ln \lambda = \alpha'' + \frac{n-1}{r}\alpha'$$

and

$$grad\,(\Delta \ln \lambda) = \left(\alpha''' + \frac{n-1}{r}\alpha'' - \frac{n-1}{r^2}\alpha'\right)\frac{\partial}{\partial r}.$$

Therefore ϕ is biharmonic if and only if the function α satisfies the following differential equation

$$\beta'' - (n-4)\,\beta\beta' + \frac{n-1}{r}\beta' - \frac{n-1}{r^2}\beta - \frac{2\,(n-1)}{r}\beta^2 - (n-2)\,\beta^3 = 0.$$

If we denote β=α′β=α′, the biharmonicity of ϕ is equivalent to the differential equation

$$\beta'' - (n-4)\,\beta\beta' + \frac{n-1}{r}\beta' - \frac{n-1}{r^2}\beta - \frac{2\,(n-1)}{r}\beta^2 - (n-2)\,\beta^3 = 0.$$

As a consequence of the Corollary 2, We will present some remarks which we give a particular solutions of the Equation (34) that allows us to construct a biharmonic non-harmonic maps.

Remark 3.

Looking for particular solutions of type β=ar (a∈R*)*. By (34), we deduce that* ϕ:(Rn,g)→(Nn,h) *(n≥3n≥3) is biharmonic if and only if a is a solution of the algebraic equation*

$$(n-2)\,a^2+(n+2)\,a+2n-2=0.$$

This equation has real solutions if and only if n∈{3,4}.

1. *If* $n=3$*, we find* $a=\frac{-5+\sqrt{17}}{2}$ *or* $a=\frac{-5-\sqrt{17}}{2}$*, so* $\lambda=Cr^{-\left(\frac{5-\sqrt{17}}{2}\right)}$ *or* $\lambda=Cr^{-\left(\frac{5+\sqrt{17}}{2}\right)}$ $(C\in\mathbb{R}_+^*)$*. It follows that any conformal map* $\phi:(\mathbb{R}^3,g)\to(N^3,h)$ *of dilation* $\lambda=Cr^{-\left(\frac{5-\sqrt{17}}{2}\right)}$ *or* $\lambda=Cr^{-\left(\frac{5+\sqrt{17}}{2}\right)}$ *is biharmonic non-harmonic.*
2. *If* $n=4$*, we find* $a=-1$ *or* $a=-2$*, so* $\lambda=\frac{C}{r^2}$ *or* $\lambda=\frac{C}{r}$ $(C\in\mathbb{R}_+^*)$*. Then, in this case any conformal map* $\phi:(\mathbb{R}^4,g)\to(N^4,h)$ *of dilation* $\lambda=\frac{C}{r^2}$ *or* $\lambda=\frac{C}{r}$ *is biharmonic non-harmonic. For example, the inversion* $\phi:(\mathbb{R}^n\backslash\{0\},g_{\mathbb{R}^n})\longrightarrow(\mathbb{R}^n\backslash\{0\},g_{\mathbb{R}^n})$ *definded by* $\phi(x)=\frac{x}{|x|^2}$ *is a conformal map with dilation* $\lambda=\frac{1}{r^2}$*. By (34), the inversion is biharmonic non-harmonic if and only if* $n=4$*.*

Remark 4.. *Looking for particular solutions of type* β $\beta=\frac{ar}{1+r^2}$ $(a\in\mathbb{R}^*)$*. By (34),* $\phi:(\mathbb{R}^n,g)\to(N^n,h)$ *is biharmonic if and only we have*

$$(n-2)\,a^2+(n+2)\,a+2n-2=0$$

and

$$3\,(n-2)\,a+2n+4=0.$$

These two equations gives a=−2a=-2 *and* n=4n=4*, it follows that the dilation is equal to* $\lambda=\frac{C}{r^2+1}$ $(C\in\mathbb{R}_+^*)$*. Then, all conformal maps* $\phi:(\mathbb{R}^4,g)\to(N^4,h)$ *of dilation* $\lambda=\frac{C}{r^2+1}$ *are biharmonic non-harmonic. For example, the inverse of the stereographic* *projection of the sphere* ϕ: $\phi : \mathbb{R}^n \longrightarrow S^n$ *definded by* $\phi(x) = \frac{1}{|x|^2+1}\left(|x|^2-1, 2x\right)$ *is a conformal map with dilation* $\lambda = \frac{2}{r^2+1}$. *By (34), the inverse of the stereographic projection is biharmonic non-harmonic if and only if* n=4.

The last part of this paper is devoted to the study of biharmonic maps between warped product manifolds, these maps were also studied in [17]. We will give some results of the biharmonicity in other particular cases.

3. BIHARMONIC MAPS AND THE WARPED PRODUCT

Let (Mm,g)Mm,g and (Nn,h)Nn,h two Riemannian manifolds and let f∈C∞(M) be a positive function. The warped product M×fN is the product manifolds M×N endowed with the Riemannian metric Gf defined, for X,Y∈Γ(T(M×N)), by

$$G_f(X,Y) = g(d\pi(X), d\pi(Y)) + (f \circ \pi)^2 h(d\eta(X), d\eta(Y)),$$

where π:M×N⟶M and η:M×N⟶N are respectively the first and the second projection. The function *f* is called the warping function of the warped product. Let X,Y∈Γ(T(M×N)), X=(X1,X2),X2, Y=(Y1,Y2).. Denote by ∇ the Levi-Civita connection on the Riemannian product M×NM×N . The Levi-Civita connection ∇˜ of the warped product M×fN is given by

$$\tilde{\nabla}_X Y = \nabla_X Y + X_1(\ln f)(0, Y_2) + Y_1(\ln f)(0, X_2) - f^2 h(X_2, Y_2)(grad \ln f, 0). \quad (35)$$

In the first, we consider a smooth map ϕ: $(M^m, g) \longrightarrow (P^p, k)$ and we defined the map ϕ˜: $\left(M^m \times_f N^n, G_f\right) \longrightarrow (P^p, k)$ by $\widetilde{\phi}(x, y) = \phi(x)$. We will study the biharmonicity of ϕ˜ϕ˜. By calculating the tension field of ϕ˜ϕ˜, we obtain the following result :

Proposition 2.

Let ϕ: ϕ:Mm,g⟶Pp,k *be a smooth map. The tension field of the map* $\left(M^m \times_f N^n, G_f\right) \longrightarrow (P^p, k)$ *defined by* $\widetilde{\phi}(x, y) = \phi(x)$ ϕ˜ *is given by*

$$\tau(\widetilde{\phi}) = \tau(\phi) + nd\phi(grad \ln f) \quad (36)$$

Proof of Proposition 2. Let us choose {ei}1≤i≤m to be an orthonormal frame on *M* and {fj}1≤j≤n to be an orthonormal frame on *N*. An orthonormal frame on M×fN is given by {(ei,0),1f(0,fj)}. Note that in this case we have dϕ˜(X,Y)=(dϕ(X),0) for any X∈Γ(TM) and Y∈Γ(TN) By definition to the tension field, we have

$$\begin{aligned}
\tau(\widetilde{\phi}) &= Tr_{G_f} \nabla d\widetilde{\phi} \\
&= \nabla^{\widetilde{\phi}}_{(e_i,0)} d\widetilde{\phi}(e_i, 0) + \frac{1}{f^2} \nabla^{\widetilde{\phi}}_{(0,f_j)} d\widetilde{\phi}(0, f_j) \\
&\quad - d\widetilde{\phi}\left(\widetilde{\nabla}_{(e_i,0)}(e_i, 0)\right) - \frac{1}{f^2} d\widetilde{\phi}\left(\widetilde{\nabla}_{(0,f_j)}(0, f_j)\right).
\end{aligned}$$

A simple calculation gives

$$\nabla^{\widetilde{\phi}}_{(e_i,0)} d\widetilde{\phi}(e_i, 0) = \nabla^{\phi}_{e_i} d\phi(e_i)$$

and

$$\nabla^{\widetilde{\phi}}_{(0,f_j)} d\widetilde{\phi}(0, f_j) = 0,$$

By using the equation (35), we deduce that

$$\widetilde{\nabla}_{(e_i,0)}(e_i, 0) = (\nabla_{e_i} e_i, 0)$$

and

$$\widetilde{\nabla}_{(0,f_j)}(0, f_j) = \left(0, \nabla_{f_j} f_j\right) - nf^2 (grad \ln f, 0).$$

It follows that

$$\tau(\widetilde{\phi}) = \nabla^{\phi}_{e_i} d\phi(e_i) - d\phi\left(\nabla^{M}_{e_i} e_i\right) + nd\phi(grad \ln f),$$

then, we obtain

$$\tau(\widetilde{\phi}) = \tau(\phi) + nd\phi(grad \ln f).$$

Remark 5.

If ϕ:(Mm,g)⟶(Pm,k) ⟶Pm,k (m≥3m≥3) *is a conformal map with dilation* λ*, the tension field of* ϕ˜ϕ˜ *is given by*

$$\tau(\widetilde{\phi}) = (2-m)\, d\phi(grad \ln \lambda) + nd\phi(grad \ln f) = d\phi\left(grad \ln\left(\lambda^{2-m} f^n\right)\right).$$

Then ϕ˜ϕ˜ *is harmonic if and only if the function* λ2−mfn *is constant.*

We will now calculate the bitension field of the map ϕ˜:(Mm×fNn,Gf)⟶(Pp,k)ϕ˜:Mm×fNn,Gf⟶Pp,k.

Theorem 4.

Let ϕ: $(M^m, g) \longrightarrow (P^p, k)$ *be a smooth map. The bitension field of the map* ϕ˜ $\left(M^m \times_f N^n, G_f\right) \longrightarrow (P^p, k)$ *defined by* ϕ˜(x,y)=ϕ(x) *is given by*

$$\tau_2(\widetilde{\phi}) = \tau_2(\phi) - n\left(Tr_g \nabla^2 d\phi\,(grad \ln f) + Tr_g R^P\left(d\phi\,(grad \ln f), d\phi\right) d\phi\right) - n\nabla_{grad \ln f}\,\tau(\phi) - n^2 \nabla_{grad \ln f}\, d\phi\,(grad \ln f). \tag{37}$$

Proof of Theorem 4. By definition of the bi-tension field, we have

$$\tau_2(\widetilde{\phi}) = -Tr_{G_f}\left(\nabla^{\widetilde{\phi}}\right)^2 \tau(\widetilde{\phi}) - Tr_{G_f} R^P\left(\tau(\widetilde{\phi}), d\widetilde{\phi}\right) d\widetilde{\phi} \tag{38}$$

For the first term $Tr_{G_f}\left(\nabla^{\widetilde{\phi}}\right)^2 \tau(\widetilde{\phi})$,˜, we have

$$Tr_{G_f}\left(\nabla^{\widetilde{\phi}}\right)^2 \tau(\widetilde{\phi}) = \nabla^{\widetilde{\phi}}_{(e_i,0)} \nabla^{\widetilde{\phi}}_{(e_i,0)} \tau(\widetilde{\phi}) + \frac{1}{f^2} \nabla^{\widetilde{\phi}}_{(0,f_j)} \nabla^{\widetilde{\phi}}_{(0,f_j)} \tau(\widetilde{\phi}) - \nabla^{\widetilde{\phi}}_{\widetilde{\nabla}_{(e_i,0)}(e_i,0)} \tau(\widetilde{\phi}) - \frac{1}{f^2} \nabla^{\widetilde{\phi}}_{\widetilde{\nabla}_{(0,f_j)}(0,f_j)} \tau(\widetilde{\phi}).$$

We will study term by term the right-hand of this expression. A simple calculation gives

$$\nabla^{\widetilde{\phi}}_{(e_i,0)}\nabla^{\widetilde{\phi}}_{(e_i,0)}\tau(\widetilde{\phi}) = \nabla^{\widetilde{\phi}}_{(e_i,0)}\nabla^{\widetilde{\phi}}_{(e_i,0)}\tau(\phi) + n\nabla^{\widetilde{\phi}}_{(e_i,0)}\nabla^{\widetilde{\phi}}_{(e_i,0)}d\phi(grad \ln f)$$
$$= \nabla^{\phi}_{e_i}\nabla^{\phi}_{e_i}\tau(\phi) + n\nabla^{\phi}_{e_i}\nabla^{\phi}_{e_i}d\phi(grad \ln f)$$

and

$$\nabla^{\widetilde{\phi}}_{(0,f_j)}\nabla^{\widetilde{\phi}}_{(0,f_j)}\tau(\widetilde{\phi}) = 0.$$

By using the equation (35), we obtain

$$\nabla^{\widetilde{\phi}}_{\widetilde{\nabla}_{(e_i,0)}(e_i,0)}\tau(\widetilde{\phi}) = \nabla^{\phi}_{\nabla^M_{e_i}e_i}\tau(\phi) + n\nabla^{\phi}_{\nabla^M_{e_i}e_i}d\phi(grad \ln f),$$

and

$$\nabla^{\widetilde{\phi}}_{\widetilde{\nabla}_{(0,f_j)}(0,f_j)}\tau(\widetilde{\phi}) = -nf^2\nabla^{\phi}_{grad \ln f}\tau(\phi) - n^2f^2\nabla^{\phi}_{grad \ln f}d\phi(grad \ln f).$$

Then, we deduce that

$$Tr_{G_f}\left(\nabla^{\widetilde{\phi}}\right)^2\tau(\widetilde{\phi}) = Tr_g\left(\nabla^{\phi}\right)^2\tau(\phi) + nTr_g\left(\nabla^{\phi}\right)^2 d\phi(grad \ln f)$$
$$+ n\nabla^{\phi}_{grad \ln f}\tau(\phi) + n^2\nabla^{\phi}_{grad \ln f}d\phi(grad \ln f). \quad (39)$$

To complete the proof, we will simplify the term $Tr_{G_f}R^P(\tau(\widetilde{\phi}), d\widetilde{\phi})\, d\widetilde{\phi}$,~, we have

$$
\begin{aligned}
Tr_{G_f} R^P \left(\tau\left(\widetilde{\phi}\right), d\widetilde{\phi}\right) d\widetilde{\phi} &= R^P \left(\tau\left(\widetilde{\phi}\right), d\widetilde{\phi}\left(e_i, 0\right)\right) d\widetilde{\phi}\left(e_i, 0\right) \\
&+ \frac{1}{f^2} R^P \left(\tau\left(\widetilde{\phi}\right), d\widetilde{\phi}\left(0, f_j\right)\right) d\widetilde{\phi}\left(0, f_j\right) \\
&= R^P \left(\tau\left(\widetilde{\phi}\right), d\widetilde{\phi}\left(e_i, 0\right)\right) d\widetilde{\phi}\left(e_i, 0\right) \\
&= R^P \left(\tau\left(\phi\right), d\phi\left(e_i\right)\right) d\phi\left(e_i\right) \\
&+ n R^P \left(d\phi\left(grad \ln f\right), d\phi\left(e_i\right)\right) d\phi\left(e_i\right).
\end{aligned}
$$

It follows that

$$
Tr_{G_f} R^P \left(\tau\left(\widetilde{\phi}\right), d\widetilde{\phi}\right) d\widetilde{\phi} = Tr_g R^P \left(\tau\left(\phi\right), d\phi\right) d\phi + n Tr_g R^P \left(d\phi\left(grad \ln f\right), d\phi\right) d\phi. \tag{40}
$$

If we replace (39) and (40) in (38), we obtain

$$
\begin{aligned}
\tau_2\left(\widetilde{\phi}\right) &= \tau_2\left(\phi\right) - n\left(Tr_g \nabla^2 d\phi\left(grad \ln f\right) + Tr_g R^P \left(d\phi\left(grad \ln f\right), d\phi\right) d\phi\right) \\
&- n\nabla_{grad \ln f} \tau\left(\phi\right) - n^2 \nabla_{grad \ln f} d\phi\left(grad \ln f\right).
\end{aligned}
$$

The proof of Theorem 4 is complete.

Remark 6.

Theorem 4 is a particular result of generalized warped product manifolds (see [18]).

As a consequence, if ϕ is harmonic, we have

Corollary 3.

Let ϕ:(Mm,g)⟶(Pp,k) *a harmonic map. The map* $\widetilde{\phi} : \left(M^m \times_f N^n, G_f\right) \longrightarrow \left(P^p, k\right)$ *defined by* ϕ˜(x,y)=ϕ(x)ϕ˜x,y=ϕx *is biharmonic if and only if*

$$Tr_g \nabla^2 d\phi \left(grad \ln f\right) + Tr_g R^P \left(d\phi \left(grad \ln f\right), d\phi\right) d\phi + n \nabla_{grad \ln f} d\phi \left(grad \ln f\right) = 0.$$

In particular if ϕ=IdM, *the first projection* $P_1 : \left(M^m \times_f N^n, G_f\right) \longrightarrow (M^m, g)$ *defined* P1x,y=x *is biharmonic if and only if (see [17])*

$$grad \Delta \ln f + \frac{n}{2} grad \left(|grad \ln f|^2\right) + 2Ricci^M \left(grad \ln f\right) = 0.$$

In the following we shall present an example of biharmonic non-harmonic maps.

Example 1.

Let $Let\ \widetilde{\varphi} : \mathbb{R}^m \setminus \{0\} \times_f N^n \longrightarrow \mathbb{R}^m \setminus \{0\}$ *defined by* $\widetilde{\varphi}(x, y) = \frac{x}{|x|^2}$ *when we suppose that* lnflnf *is radial* (lnf=α(r)). *Then by Theorem 4, we deduce that the map* $\widetilde{\varphi} : \mathbb{R}^m \setminus \{0\} \times_f N^n \longrightarrow \mathbb{R}^m \setminus \{0\}$ *is biharmonic if and only if the function α satisfies the following differential equation*

$$n\alpha''' + \frac{n(m-5)}{r}\alpha'' - \frac{3n(3m-7)}{r^2}\alpha' + n^2\alpha'\alpha'' - \frac{2n^2}{r}(\alpha')^2 - \frac{8(m-2)(m-4)}{r^3} = 0.$$

Let β=α′β=α′, *this equation becomes*

$$n\beta'' + \frac{n(m-5)}{r}\beta' - \frac{3n(3m-7)}{r^2}\beta + n^2\beta\beta' - \frac{2n^2}{r}\beta^2 - \frac{8(m-2)(m-4)}{r^3} = 0.$$

Looking for particular solutions of type β=ar *(a∈R∗), then* $\beta = \frac{a}{r}\ (a \in \mathbb{R}^*)$, *then* $\widetilde{\varphi} : \mathbb{R}^m \setminus \{0\} \times_f N^n \longrightarrow \mathbb{R}^m \setminus \{0\}$ *is biharmonic if and only if*

$$3n^2a^2 + 2n\,(5m - 14)\,a + 8\,(m-2)\,(m-4) = 0.$$

This equation has two solutions $a = \frac{4-2m}{n}$ *and* $a = \frac{4(4-m)}{3n}$.

1.

1. *For* $a = \frac{4-2m}{n}$, *we obtain* $f(r) = Cr^{\frac{4-2m}{n}}$ *and in this case* $\widetilde{\varphi} : \mathbb{R}^m \setminus \{0\} \times_f N^n \longrightarrow \mathbb{R}^m \setminus \{0\}$ *is harmonic so biharmonic.*
2. *For* $a = \frac{4(4-m)}{3n}$, *we obtain* $f(r) = Cr^{\frac{4(4-m)}{3n}}$ *and in this case* $\widetilde{\varphi} : \mathbb{R}^m \setminus \{0\} \times_f N^n \longrightarrow \mathbb{R}^m \setminus \{0\}$ *is biharmonic non-harmonic.*

Now, we consider a smooth map ψ:(Nn,g)⟶(Pp,k) and we define the map ψ˜: $\left(M^m \times_f N^n, G_f\right) \longrightarrow (P^p, k)$ by $\widetilde{\psi}(x,y) = \psi(y)$. We will study the biharmonicity of ψ˜ψ˜, we obtain the following result :

Theorem 5.

Let ψ:(Nn,h)→(Pp,k) *be a smooth map, we define* $\widetilde{\psi} : \left(M^m \times_{f^2} N^n, G_{f^2}\right) \rightarrow (P^p, k)$. *The tension field and the bi-tension field of* ψ˜ *are given by*

$$\tau\left(\widetilde{\psi}\right) = \frac{1}{f^2 \circ \pi}\,\tau(\psi) \tag{41}$$

And

$$\tau_2(\tilde{\psi}) = \frac{1}{f^4 \circ \pi}\tau_2(\psi) - \frac{2}{f^2 \circ \pi}\left(\left(\Delta \ln f + (n-2)|grad \ln f|^2\right) \circ \pi\right)\tau(\psi).$$

(42)

Proof of Theorem 5. In the first, we calculate the tension field of of ψ˜ψ˜. By definition to the tension field, we have

$$\tau(\tilde{\psi}) = Tr_{G_f} \nabla d\tilde{\psi}$$
$$= \nabla^{\tilde{\psi}}_{(e_i,0)} d\tilde{\psi}(e_i,0) + \frac{1}{f^2 \circ \pi}\nabla^{\tilde{\psi}}_{(0,f_j)} d\tilde{\psi}(0,f_j)$$
$$- d\tilde{\psi}\left(\tilde{\nabla}_{(e_i,0)}(e_i,0)\right) - \frac{1}{f^2 \circ \pi} d\tilde{\psi}\left(\tilde{\nabla}_{(0,f_j)}(0,f_j)\right).$$

By using the equation (35), we obtain

$$\tau(\tilde{\psi}) = \frac{1}{f^2 \circ \pi}\nabla^{\psi}_{f_j} d\psi(f_j) - \frac{1}{f^2 \circ \pi} d\psi\left(\nabla_{f_j} f_j\right) = \frac{1}{f^2 \circ \pi}\tau(\psi),$$

then

$$\tau(\tilde{\psi}) = \frac{1}{f^2 \circ \pi}\tau(\psi).$$

By this expression, we deduce that ψ˜ψ˜ is harmonic if and only if ψ is harmonic. Now, we will calculate the bi-tension field of ψ˜. By definition, we have

$$\tau_2(\tilde{\psi}) = -Tr_{G_f}\left(\nabla^{\tilde{\psi}}\right)^2 \tau(\tilde{\psi}) - Tr_{G_f} R^P\left(\tau(\tilde{\psi}), d\tilde{\psi}\right) d\tilde{\psi}.$$

(43)

For the first term $Tr_{G_f}\left(\nabla^{\tilde{\psi}}\right)^2 \tau(\tilde{\psi})$,˜, we have

$$Tr_{G_f}\left(\nabla^{\widetilde{\psi}}\right)^2 \tau\left(\widetilde{\psi}\right) = \nabla^{\widetilde{\psi}}_{(e_i,0)}\nabla^{\widetilde{\psi}}_{(e_i,0)}\tau\left(\widetilde{\psi}\right) + \frac{1}{f^2\circ\pi}\nabla^{\widetilde{\psi}}_{(0,f_j)}\nabla^{\widetilde{\psi}}_{(0,f_j)}\tau\left(\widetilde{\psi}\right)$$
$$-\nabla^{\widetilde{\psi}}_{\widetilde{\nabla}_{(e_i,0)}(e_i,0)}\tau\left(\widetilde{\psi}\right) - \frac{1}{f^2\circ\pi}\nabla^{\widetilde{\psi}}_{\widetilde{\nabla}_{(0,f_j)}(0,f_j)}\tau\left(\widetilde{\psi}\right).$$

A long calculation gives

$$\nabla^{\widetilde{\psi}}_{(e_i,0)}\nabla^{\widetilde{\psi}}_{(e_i,0)}\tau\left(\widetilde{\psi}\right) = \frac{2}{f^2\circ\pi}\left(\left(2\left|grad\ln f\right|^2 - e_i\left(e_i\left(\ln f\right)\right)\right)\circ\pi\right)\tau(\psi)$$

and

$$\frac{1}{f^2\circ\pi}\nabla^{\widetilde{\psi}}_{(0,f_j)}\nabla^{\widetilde{\psi}}_{(0,f_j)}\tau\left(\widetilde{\psi}\right) = \frac{1}{f^4\circ\pi}\nabla^{\psi}_{f_j}\nabla^{\psi}_{f_j}\tau(\psi).$$

Finally, by (35), we obtain

$$\nabla^{\widetilde{\psi}}_{\widetilde{\nabla}_{(e_i,0)}(e_i,0)}\tau\left(\widetilde{\psi}\right) = \frac{2}{f^2\circ\pi}\left(\nabla_{e_i}e_i\left(\left(\ln f\right)\right)\circ\pi\right)\tau(\psi)$$

and

$$\frac{1}{f^2\circ\pi}\nabla^{\widetilde{\psi}}_{\widetilde{\nabla}_{(0,f_j)}(0,f_j)}\tau\left(\widetilde{\psi}\right) = \frac{1}{f^4\circ\pi}\nabla^{\psi}_{\nabla_{f_j}f_j}\tau(\psi) + \frac{2n}{f^2\circ\pi}\left(\left(\left|grad\ln f\right|^2\right)\circ\pi\right)\tau(\psi).$$

Which gives us

$$Tr_{G_f}\left(\nabla^{\widetilde{\psi}}\right)^2\tau\left(\widetilde{\psi}\right) = \frac{1}{f^4\circ\pi}Tr_h\nabla^2\tau(\psi) - \frac{2}{f^2\circ\pi}\left(\left(\Delta\ln f + (n-2)\left|grad\ln f\right|^2\right)\circ\pi\right)\tau(\psi) \quad (44)$$

Finally for the first term $Tr_{G_f}R^P\left(\tau\left(\widetilde{\psi}\right), d\widetilde{\psi}\right)d\widetilde{\psi}$, it is easy to verify that

$$Tr_{G_f} R^P \left(\tau\left(\widetilde{\psi}\right), d\widetilde{\psi}\right) d\widetilde{\psi} = \frac{1}{f^4 \circ \pi} Tr_h R^P \left(\tau\left(\psi\right), d\psi\right) d\psi. \quad (45)$$

If we substitute (44) and (45) in (43), we obtain

$$\tau_2\left(\widetilde{\psi}\right) = \frac{1}{f^4 \circ \pi}\tau_2\left(\psi\right) - \frac{2}{f^2 \circ \pi}\left(\left(\Delta \ln f + (n-2)\left|grad \ln f\right|^2\right) \circ \pi\right)\tau\left(\psi\right).$$

This completes the proof of Theorem 5. An immediate consequence of Theorem 5 is given by the following corollary :

Corollary 4.

Let $\psi : (N^n, h) \longrightarrow (P^p, k)$ *a biharmonic non-harmonic map. The map* $\phi\tilde{}\left(M^m \times_f N^n, G_{f^2}\right) \longrightarrow (P^p, k)$ *defined by* $\widetilde{\psi}(x,y) = \psi(y)$ *,kdefined by* $\widetilde{\psi}(x,y) = \psi(y)$ *is biharmonic if and only if the function* fn−2fn-2 *is harmonic.*

ACKNOWLEDGMENTS

The authors would like to thank the referee for some useful comments and their helpful suggestions that have improved the quality of this paper.

AUTHOR CONTRIBUTIONS

The authors provide equal contributions to this paper.

CONFLICTS OF INTEREST

The authors declare no conflict of interest.

REFERENCES

1. Eells, J.; Lemaire, L. A report on harmonic maps. *Bull. Lond. Math. Soc.* **1978**, *16*, 1–68.

2. Eells, J.; Lemaire, L. Another report on harmonic maps. *Bull. Lond. Math. Soc.* **1988**, *20*, 385–524.

3. Eells, J.; Lemaire, L. *Selected topics in harmonic maps*; CBMS Regional Conference Series in Mathematics; American Mathematical Society: Providence, RI, USA, 1981; Vol. 150.

4. Eells, J.; Ratto, A. *Harmonic Maps and Minimal Immersions with Symmetries: Methods of Ordinary Differential Equations Applied to Elliptic Variational Problems*; Princeton University Press: Princeton, NJ, USA, 1993; Vol. 130.

5. Baird, P.; Eells, J. *A Conservation Law for Harmonic Maps*; Lecture Notes in Math. 894; Springer: Berlin, Germany, 1981; pp. 1–25.

6. Jiang, G.Y. 2-harmonic maps and their first and second variational formulas. *Chin. Ann. Math. Ser.* **1986**, *A7*, 389–402.

7. Loubeau, E.; Montaldo, S.; Oniciuc, C. The stress-energy tensor for biharmonic maps. *Math. Z.* **2008**, *259*, 503–524.

8. Baird, P.; Kamissoko, D. On constructing biharmonic maps and metrics. *Ann. Glob. Anal. Geom.* **2003**, *23*, 65–75.

9. Balmus, A. Biharmonic properties and conformal changes. *Analele Stiintifice ale Univ. Al.I. Cuza Iasi Mat.* **2004**, *50*, 367–372.

10. Ou, Y.-L. p-harmonic morphisms, biharmonic morphisms, and non-harmonic biharmonic maps. *J. Geom. Phys.* **2006**,*56*, 358–374.

11. Ouakkas, S. Biharmonic maps, conformal deformations and the Hopf maps. *Diff. Geom. Appl.* **2008**, *26*, 495–502.

12. Baird, P.; Fardoun, A.; Ouakkas, S. Conformal and semi-conformal biharmonic maps. *Ann. Glob. Anal. Geom.* **2008**, *34*, 403–414.

13. Montaldo, S.; Oniciuc, C. A short survey of biharmonic maps between Riemannian manifolds. *Rev. Union Mat. Argent.***2006**, *47*, 1–22.

14. Baird, P. *Harmonic Maps with Symmetry, Harmonic Morphisms and Deformation of Metrics*; Research Notes in Mathematics; CRC Press: London, UK, 1983; pp. 27–39.

15. Baird, P.; Loubeau, E.; Oniciuc, C. Harmonic and biharmonic maps from surfaces. In *Harmonic maps and differential geometry*; Amer. Math. Soc.: Providence, RI, USA, 2011; pp. 234–241.

16. Baird, P.; Wood, J.C. *Harmonic Morphisms between Riemannain Manifolds*; London mathematical Society Monographs (N.S.); Oxford University Press: Oxford, UK, 2003.

17. Balmus, A.; Montaldo, S.; Oniciuc, C. Biharmonic maps between warped product manifolds. *J. Geom. Phys.* **2007**, *57*, 449–466.

18. Djaa, N.E.H.; Boulal, A.; Zagane, A. Generalized warped product manifolds and biharmonic maps. *Acta Math. Univ. Comen.* **2012**, *81*, 283–298.

Index